EPIC
HOMESTEADING

Quarto.com

© 2024 Quarto Publishing Group USA Inc.
Text and Photography © 2024 Kevin Espiritu

First Published in 2024 by Cool Springs Press, an imprint of The Quarto Group,
100 Cummings Center, Suite 265-D, Beverly, MA 01915, USA.
T (978) 282-9590 F (978) 283-2742

Cool Springs Press titles are also available at discount for retail, wholesale, promotional, and bulk purchase. For details, contact the Special Sales Manager by email at specialsales@quarto.com or by mail at The Quarto Group, Attn: Special Sales Manager, 100 Cummings Center, Suite 265-D, Beverly, MA 01915, USA.

28 27 26 25 24   1 2 3 4 5

ISBN: 978-0-7603-8376-6

Digital edition published in 2024
eISBN: 978-0-7603-8377-3

Library of Congress Cataloging-in-Publication Data available

Design: Samantha J. Bednarek, samanthabednarek.com
Cover Image: Ian Ware
Page Layout: Samantha J. Bednarek, samanthabednarek.com
Photography: Ian Ware and Kevin Espiritu, except page 22 by Sara Bendrick; pages 123 (bottom), 124, 125 (top) by Steve Churchill; page 89 by John Davis Hunks; page 68 by Gardyn; page 136 by Google Project Sunroof; page 66 by Angelica LaVallee; pages 14, 62, 64, 65, 95, 115 (bottom), 156 (bottom), 173 (top), 174, 183, 184, 186 (middle and bottom) by Shutterstock; page 115 (top) by Yimby Compost.
Illustration: Zoe Naylor except small icons and page 64 by Shutterstock and radish logo by Epic Gardening

Printed in China

EPIC
HOMESTEADING

Your Guide to Self-Sufficiency
on a Modern, High-Tech,
Backyard Homestead

KEVIN ESPIRITU
OF EPIC GARDENING

with contributing editor Lisa Munniksma

COOL
SPRINGS
PRESS

CONTENTS

The Epic Homestead from above. Replete with solar panels, rainwater capture, and a productive garden, all in a typical suburban neighborhood.

INTRODUCTION

I started homesteading in search of more freedom in my life without sacrificing modern comforts. The 0.3-acre (0.12 ha) lot in USDA Hardiness Zone 10b—San Diego—that I call the Epic Homestead looked *very* different at the beginning of my journey than it does as I write this book. In just a few seasons, I've taken this nearly bare urban lot and turned it into a thriving homestead.

When I moved in, the property had seven different kinds of fencing and a few fruit and nut trees, and the yard was nearly covered in foxtail barley. I had a 1,000-square-foot (93 m²) house with the typical structures, a small shed in the back, a garage, and a covered patio. This was not what most people think of when they think of an ideal slate for self-sufficiency, but in this space, I saw the potential to build an efficient, modern homestead.

As I'm writing this book, the homestead now has:
- 17 raised bed gardens with automated irrigation in the front yard
- A 5,700-gallon (21,577 L) water-catchment system,
- A new roof with 25 solar panels
- An outdoor shower with greywater running to the citrus orchard
- And a whole pond ecosystem and gorgeous patio area

I'm starting to replace sections of the mismatched fence to be more uniform and also to take the fence to the actual edge of the property, maximizing my space for growing. I've filled out a citrus hedge, planted passion fruit and avocado, and—no surprise to anyone who knows my favorite fruit—built an area that I like to call Dragon Fruit Alley. There's also the Epic mural by my friends Nicholas Danger and Skye Walker, featuring our garden hermit Jacques, a couple of chickens, a patch of rich soil, a koi, Babka the cat, some vegetables, and, of course, a dragon fruit. There's so much more that's happened on my homestead in just three years, and I'm excited to share it all with you in this book, just as I have been on the Epic Gardening and Epic Homesteading channels.

My suburban lot is about 13,000 square feet (1,208 m²), or about 0.3 acres (0.12 ha). I had to get creative to fit all of the elements of a productive, sustainable homestead on the property. Read on to learn how I did it.

Why Homestead Now?

We all come into homesteading for different reasons, and more people are finding more reasons to start their path toward some level of self-sufficiency. There are three important reasons that I'm interested in homesteading.

A secure food supply

Our global, industrialized food system runs on a two-week, just-in-time supply chain. Without peeking behind the scenes of worldwide food production and distribution, you might not realize that we're really just three days away from chaos if those delivery trucks stop running. In fact, for the past sixty or more years, most of us have been so far removed from where our food comes from that we've never even thought about what would happen if grocery store shelves weren't stocked. We started to see the reality of this in 2020, in the early days of the COVID-19 pandemic.

The "front yard grocery store"—a dozen raised beds that provide most of the produce I eat.

I'm interested in homesteading from a self-sufficiency angle, thinking about the position I want to be in if the food-system supply chain goes bad. Call it food insurance. Most of us buy life insurance or health insurance, paying hundreds of dollars a month in case of emergency. Stocking up on two months' worth of water, health supplies, and food—even canned goods—may be the cheapest insurance you can buy if something in the food system really did go poorly.

When homesteading, you don't have to fill your pantry with food from the grocery store. You can learn how to actually produce that food yourself. Now you have the practical angle of self-sufficiency, then you tie in the psychological and emotional angle of just how *good* it feels to be able to grow something. You're no longer just a consumer; you're becoming a *producer*, which inherently feels good.

Environmental concerns

The industrial food system has done little to improve the environment over the years. Vast expanses of chemically controlled crops grown in monoculture, food is grown on one continent and shipped to another, nutrients run off in waterways, and the proliferation of antibiotic-resistant bacteria are just a handful of environmental costs that concern me in the current food system. There's also extreme waste in all areas of the industrial food system with as much as 40 percent of purchased produce being wasted—plus the produce that doesn't even make it to the store. We're using resources upstream to produce waste downstream, and it doesn't make sense.

Tallying up the environmental costs of the food system, it would be easy enough to say our individual actions don't matter. We're seeing environmental degradation, severe droughts, and more unpredictable weather all the time. But I firmly believe that any ecosystem improvement you can make locally does help globally.

We can only control what we can control, and in this case, we can control what goes in our yards, our gardens, and our mouths. I'm not going to tell you that you should compost, grow and preserve your own food, install solar panels, and reuse greywater because of the environment. I'm going to continue to do what I've done all along and let you find the philosophy that helps you transform your garden and your life.

Food quality

This is not the book for me to get into conventional farming versus USDA certified organic farming. What I can say is that fruits and vegetables grown in healthy soil are more nutrient dense than those grown in soil that's been stripped of its nutrients. Scientists are finding that foods grown in soils containing all the major nutrients that plants need often have higher yields and higher nutrient concentrations. You can read all about this in study after study from the University of Washington, University of Hohenheim, Makerere University, Justus Liebig University Giessen, the World Agroforestry Centre, and others.

Multiple studies also show that on average, organic crops had one-fifth to almost one-third more vitamin C, iron, and magnesium; higher overall levels of mineral micronutrients; and lower levels of nitrate and heavy metals.

In conventionally cropped soils, especially on land that has been farmed for decades, naturally occurring nutrients may have been farmed out, and they are being replaced with synthetic fertilizers that may not replace all the nutrients we need—especially not the micronutrients. The chemicals are not a friend to the soil life, either.

The nutrient value of fruits and vegetables is highest at harvest time. Produce spends anywhere from a few days to a few weeks in transit from the farms where it's grown to the stores where it's purchased. All the while, the nutritional value of these foods is decreasing. The most nutrition comes from the freshest produce, like the produce you've harvested yourself from your yard or container garden.

And do I need to mention taste? Food that's harvested 1,000 or more miles (1,600 or more km) from where it's eaten has to be picked before it's ripe, so it won't spoil during transit. The mainstream food system doesn't choose fruit and vegetable varieties based on whether they taste good but on their appearance and ability to keep the longest. By growing my own produce, I also get the best taste and quality.

Modern Homesteading

I'm not interested in slogging and suffering my way through this homesteading journey. I'm trying to add as much modern technology that makes sense without losing the fact that we should know where our food comes from and ideally produce a significant chunk of it ourselves.

The word "homesteading" conjures different images for different people. You may be picturing a pastoral setting; a milk cow in the background; and a pantry full of canned fruits, vegetables, and meats to last through the year. While this is a common homesteading vision, it wasn't mine. Consider, instead, the idea of high-tech, natural living.

Homo sapiens evolved outdoors, in a community. A lot of people have a longing to return to those roots now. I am one of them, but instead of feeling a pull to do things the traditional way, I see a bigger vision of how we can use modern conveniences to enhance our homesteading experience.

The artichoke patch, watered by a combination of laundry grey water and captured rain.

HOW TO USE THIS BOOK

If you want to learn how to successfully homestead but you think you don't have enough space or enough time to dedicate to it, I wrote this book for you. I knew nothing about homesteading when I first started, and I sure didn't expect I'd be making a living sharing my homesteading journey with you. Yet here I am, living the homesteading life on 0.3 acre (0.12 ha) in San Diego.

I'm often asked how I've gotten this far in my own homesteading venture. I've invested time in reading books, watching YouTube channels, learning from friends, and just trying things out. In this book, you'll find gardening tips and tricks, easy DIY energy- and water-conservation projects, chicken-care advice, some of my favorite recipes, and more photos and information from the Epic Homestead.

Each chapter has a QR code that takes you to the Epic Homesteading webpage dedicated to that chapter. There you'll find videos, articles, and more projects related to what you'll learn in that chapter.

I'm aware that my projects tend to be "epic-sized," and because of that, I intend for you to take what you need from this book. Don't think you have to follow each project to the T or commit to memory everything from each chapter. There is no standard right way to do everything on every homestead. Being a successful modern homesteader is more about shifting your mindset from consumer to producer than it is about doing big, fancy projects. Throughout the book, I include ways that you can scale up or scale down the work I've done here.

I'm sharing in these pages what made sense for me and worked in my space. Take and use what makes sense for you. If you like the idea of a project but think you want to tweak some things, take that idea and make it your own. If your yard is half the size of my yard, or if you don't have a yard at all, you can still learn from my homesteading successes (and nonsuccesses) for yourself. There's something in this book for homesteaders of all experience levels and areas of interest.

Scan for video

Your homesteading journey is yours, but I wish I had this book when starting out on my own path.

While our culture has changed to put many of us at desks doing largely solitary work most of the time, the bodies we're in still respond in physical, psychological, and spiritual ways to interacting with the natural world and being around family and friends. These opportunities are inherent in homesteading.

Being in tighter-knit cultures can improve your happiness, and a feeling of connectedness tends to happen more when you are a homesteader. I think about this when I'm sharing food, whether I'm donating food to the San Diego Food Bank, giving eggs to my mom, or enjoying a meal with someone, it just feels more wholesome. It feels like a better way to live.

From a modern perspective, technology can boost most things we do on a homestead. Our ability to learn about homesteading has never been greater than it is today with the internet and all the platforms and creators that exist. Instead of watering gardens by hand, technology allowed me to install a pretty robust internet-connected drip-irrigation system. You can set up all kinds of tools on timers, use rainwater-capture systems and solar energy, and put in time with a planning app to make your crop production a little more efficient.

Modern homesteading means growing more of our own food, eating better-quality food, using fewer synthetic chemical inputs into our lives, having more time outside, and getting more exercise. High-tech, natural living improves upon the work and systems that our ancestors developed to care for themselves for centuries. I challenge myself to use the best of both worlds.

While I've built the Epic Homestead for food production and efficiency, I don't think I'll ever be 100 percent self-sufficient. On 0.3 acre (0.12 ha) in San Diego, it would be hard to raise all my own proteins. And while I strive to eat seasonally from the garden, every now and then, I will purchase an out-of-season avocado or a tomato from the grocery store if I really want it.

Trending toward 50 percent self-sufficient would be ideal for me, with my lifestyle, in this space. You can find your own ideal and build your homestead to suit that goal, and you don't have to give up a bit of modern conveniences to get there.

MEET JACQUES

If you spend any time on the Epic Gardening and Epic Homesteading YouTube channel and social media, you know there's a whole team of people whose expertise I admire. Throughout this book, there's one person who comes up again and again. Here, I'd like to introduce you to Jacques Lyakov.

Jacques, The Garden Hermit, my original garden assistant—this talented and knowledgeable gardener goes by many names. After going to graduate school for geology, Jacques turned his research skills toward growing food and found his true passion. He's a San Diego urban gardener like me, and his backyard garden is itself epic.

Find him on social media as Jacques In The Garden, where he focuses on flavor, low-cost DIY options, and low soil-disturbance methods.

Where you decide to build your homestead is the single most important decision you'll make on your journey. While it's possible anywhere, there are some key factors to consider. Zoning, space requirements, and climate are just some of many.

chapter one

SITE SELECTION

I HAD BEEN LOOKING FOR HOUSES for four or five months when I found the property that is now the Epic Homestead. As soon as I saw the 0.3-acre (0.12 ha) lot it sits on, I knew this was the one I wanted. It had already-producing fruit trees, a small house, and flat ground. The yard is mostly sunny, and because it's just a few miles from the coast, it has a coastal breeze coming through to knock out the staggering heat I'd find farther inland. Also, being in an urban setting, any improvements I make to this property will improve its value.

These may not be typical considerations for homesteading. We're each in this for our own reasons, and whatever your reasons are will help you determine whether a property is right for you. If you're interested in homesteading in a rural area, you're probably going to want more than 0.3 acre (0.12 ha). For an urban or suburban homestead, I think this is a perfect size.

In the United States, each municipality has its own way of going about planning and zoning. There's often a comprehensive land-use plan that outlines the type of development allowed in each area of a city or county. There's usually an elected or government-appointed board that oversees how the comprehensive plan is carried out. That board enforces regulations to be sure new development and activities are compatible with existing uses while separating "incompatible" land uses.

Already you might understand why zoning can be a challenge for homesteaders. People who don't homestead don't typically understand the inner workings of home-steading. If you've never lived alongside chickens, you don't know they're quieter than dogs. If you've always had a green lawn, you might have trouble seeing the beauty in a front-yard herb garden. Unfortunately, a lot of these zoning ordinances are designed by people who aren't homesteaders.

ZONING

Because I'm an urban homesteader, I want to focus here on a major issue that those of us in the cities and suburbs must think about: zoning. Even rural homesteaders have to deal with planning and zoning regulations, but these tend to be less strict or more well-suited to a homesteading lifestyle than urban and suburban regulations.

 Scan for video

It's important to understand the zoning regulations in your area before you settle in. Some that impact homesteading might include:

- Whether you can keep chickens, how many you can keep, if any can be roosters, the type of enclosure they require, and how far from the property line they must be kept.
- Nuisance laws that allow neighbors to report activity they see as offensive or interfering with their enjoyment of their property. The definition of "nuisance" varies, and it's typically broad.
- The size of the garden shed you're allowed to build without needing a building permit and inspection.
- Permissions required to have a garden on a rooftop.

Some places have right-to-farm laws that protect us from being accused of creating a nuisance and allow reasonable homesteading activities to take place. These are usually found in rural areas and in the boundaries of urban, suburban, and rural areas.

The concept of community is central to homesteading, and many zoning headaches can be avoided with a little community building. Your property won't break nuisance laws if it's not reported as a nuisance. Talk to your neighbors before you make huge changes. Like I said, people who don't homestead typically don't understand homesteading. Give the people around you a chance to understand that your bees won't bother them, and promise a jar of honey. Show your neighbors photos of the garden you're hoping to build, and let them get excited about receiving a few extra tomatoes now and then.

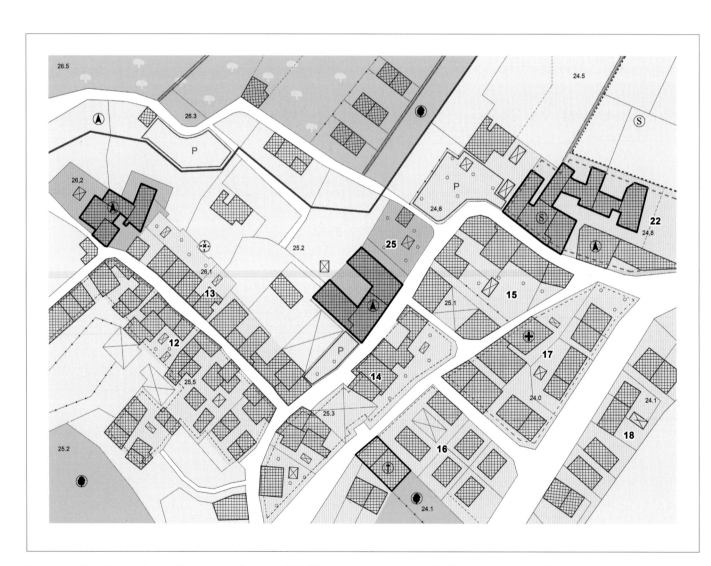

Zoning and local regulations will determine the potential of the homestead you can build—do your research!

If you can help it, I would suggest avoiding homeowners associations (HOAs), as they have an additional layer of property-use regulations. I know there are some good ones out there, but by and large, HOAs impose restrictions that make the neighborhood unsuitable for homesteading. If you're already in a neighborhood with an HOA, look into the covenants, conditions, and restrictions (CC&Rs) that describe what you can and cannot do with your property and the penalties you'd face if you did those things. HOAs can change their rules at any time without any input from you, so if there were no rules saying you can't have solar panels before you put them up, a rule could very well be imposed if someone in the neighborhood doesn't like your solar panels.

YOUR PERFECT CLIMATE

Homesteading in a coastal desert climate is pretty ideal for me, yet people manage to homestead successfully and with great enjoyment the world over, in all climates. When gardeners here in the United States talk about their climate, the U.S. Department of Agriculture (USDA) Plant Hardiness Zones are usually the first descriptor they mention. Plant hardiness zones give a quick, ballpark idea of a growing season.

In 1960, the USDA split up the country into zones according to the average annual minimum temperature. USDA Zone 1a—the interior of Alaska—is the coldest; Zone 13b—Hawaii—is the warmest. The Epic Homestead is in Zone 10b. Other countries have their own plant hardiness zones that can be translated to the USDA zones for comparison. Look at most any seed packet or plant tag, and you will find the USDA zone listed for that plant. Some seed and plant websites even allow you to search for the right varieties for your property using your hardiness zone. This is especially true for perennial plants. One example is lemongrass, which is a perennial in my garden, but if I were in Zone 6b, I would have to grow it as an annual.

If I lived in an area with an HOA, I'd probably run into some trouble with these front-yard raised beds.

While the USDA Plant Hardiness Zones are a handy, broad overview of what you can grow in your area, they don't hold all the answers to your garden capabilities. To start, the USDA Hardiness Zones aren't updated regularly enough to keep up with the changing climate. It's confusing because the temperature rated for the hardiness zone you're in now might not line up with what you've experienced living there. Even more significant than that, zones overlook the average high temperature in your area and average rainfall, first- and last-frost dates, and microclimates. Think beyond the hardiness zone for a holistic picture of what will succeed in your homestead garden.

GET YOUR ZONE

Find your USDA Plant Hardiness Zone online at
https://planthardiness.ars.usda.gov.

My 1,500 gallon (5,678 L) backyard pond creates an entire mini ecosystem of its own.

Running water attracts plenty of insects and animals.

Microclimates

Within the same zone—even on the same block or within the same county—homesteaders can experience different growing conditions thanks to microclimates. These areas have a climate that differs from the areas around them, sometimes by as much as one whole hardiness zone.

Here are a few examples of microclimates:
- A spot in your garden shaded by trees or buildings around you
- A dark painted wall or solid fence that will absorb heat from the sun and throw the heat back onto the plants
- A garden in a valley, where mornings and evenings will be colder and frost is more likely to form
- A garden on a hill, where more wind will blow, and the days will be hotter and sunnier for a longer period

Each property has its own microclimates, and you can learn to use yours to your advantage.

BUILD YOUR OWN MICROCLIMATE

Rather than being confined to the climate you were given, my friend David the Good, also known as The Survival Gardener, encourages gardeners to create microclimates to help support what you want to grow.

Gain as Much as One Zone

- Plant 3 to 4 feet (0.9 to 1.2 m) away from a south-facing wall. This wall can reflect a radiant heat of 5°F to 10°F °F (3°C to 6°C).
- Move your containers into sunnier spaces as summer turns to fall.
- Use a simple season extender, like a cloche or floating row cover, to trap the heat from the soil and keep it closer to your plants. Go epic and build a cold frame or low tunnel for more sturdy frost protection.
- Use a dark-colored mulch, like wood chips, to insulate the garden soil, warm it faster in the spring, and keep it warmer in the fall.
- In a container garden, group pots together so they protect one another instead of allowing them to be exposed to colder air circulating all around them.

Knock Off a Zone

- Plant in an area that gets some shade throughout the day. The temperature will naturally be lower.
- Create a simple shade structure using PVC and shade cloth, or use hoops for your raised bed to create a shade low tunnel. Shade cloth is a permeable woven or knitted polyethylene or polypropylene fabric. You'll find it in varying densities or degrees of shade. A 50 percent shade cloth will allow 50 percent of the sun's rays through, cutting the heat to your plant in half. Shade cloth not only keeps the plant cooler but also keeps the soil cooler, leading to less moisture evaporation.
- Use 2 or 3 inches (5 or 8 cm) of a light-colored mulch layer, like straw, which will reflect heat away from the soil rather than absorb it.

Grouping grow bags together creates an insulated air layer that can boost your effective growing zone by zone.

Frost Dates

Knowing your first and last expected frost dates Is essential to understanding your gardening capacity. At the Epic Homestead in San Diego, I rarely see frosts while homesteaders in most other places experience frost and freezes that drastically change, if not end, their gardening season. Gardeners in the south obviously have a first frost date that's later than gardeners in the north, but this can vary by weeks or more if you're in the mountains or on the coast—even within the same hardiness zone. Knowing the length of your growing season will help you select vegetable and fruit varieties that should mature in that window of time.

Average Rainfall

Water is essential enough to a garden that there's a whole chapter about it in this book (see page 147). Before setting up a homestead, have an idea of how much water to expect. When you live somewhere with a lot of rain, you have abundant opportunities to capture that water, and you also may have to mitigate too much water on your property with thoughtful design. With very little rain, you can figure out how to keep and use as much as possible and know to plant varieties that tolerate less water.

Average High Temperature

Those high temps can sneak up on you in the summer. Parts of the United States can see winter weather below 0°F (-18°C) and summer weather above 100°F (38°C). Your area's high temperature patterns are important for your garden and also for your livestock or mini livestock, like chickens. Like some plant varieties, some chicken breeds are better suited for hot versus cold weather.

A rare hailstorm at my homestead; the first in over a decade.

YOUR FIRST-FROST AND LAST-FROST DATES

Find your first-frost and last-frost dates from The National Gardening Association:
www.garden.org/apps/frost-dates

Before digging out the pond, I had to make sure I didn't hit the already-installed cistern water pipe.

Digging out the pond caused me to rebuild a fence to get the machinery in the backyard.

PROJECT PLANNING

Once you have your location, the fun part begins. It's time to homestead. There are two ways to approach your new project: run it and gun it, or take time and plan. I have been known to jump right into things, but I am trying to build a modern homestead that stands the test of time. That means I'm thoughtfully planning and hoping that I'm doing things the right way up front.

If I could redo some of the projects I did to set up the Epic Homestead, I would have done a few things in a different order. I would have figured out the fencing and security first. The property had a hodgepodge of fencing types, and I knew when I bought the property that I would need to replace some of them. I should have done that first, but I put in the shed just 5 feet (1.5 m) from the fence line before I updated the fence, making the fence-installation project a little frustrating. I also would have installed the whole solar project at once to save money and time, instead of having two separate projects.

I wish I had run irrigation and buried electrical and water lines before putting in other infrastructure, but I can see this both ways. On the one hand, the design for the garden irrigation lines evolved over my first year here, so if I had put those in first, they probably wouldn't have mapped out to the exact place I need them now. That makes me think I did this in the right order. On the other hand, I already had some hardscaping in place when it came time to run the irrigation lines, so I had to work around obstacles come trenching time. This makes me think I should have swapped the order of things. Maybe there is no perfect order of projects here and part of the homesteading journey is a few bumps along the way.

Pick Your Projects

Consider the big projects you want to do and in what order you want to do them. Your homesteading priorities, your budget, and your needs will dictate this for you. For me, that meant replacing the garden shed before tackling other projects because I needed a place to store my tools so I could more efficiently put in the rest of the garden. My compost system, on the other hand, came in much later because I knew it was important but saw it as offering a lower return.

List all of the projects that you want to do, whether you're doing them yourself or contracting them out, the time you expect to invest, and an estimated cost. I like to use the note-taking app Notion to organize my projects.

Know Your Limitations

I would prefer to build everything myself, but having little construction experience, I can admit when a project is too big for me to complete. A compost bin? No problem. A 10-by-12-foot (3 x 3.6 m) garden shed? Someone else can handle that job.

If you're going to do a project, do it right the first time. Consult with the best person in your area that you can find. Oftentimes, the "best person" means the one who is going to show up when they're supposed to show up, do quality work, and communicate well.

Practice Patience

There are a few homestead projects that I encourage you to wait on. One is bees. I'll get into this more in Chapter 8, but the summary is that bees like stability and quiet, and homestead construction is not that.

Observe

Don't make any huge moves without knowing how the light falls, how the water moves, and other nuances of your property. You might also want to test things out for a sense of how much it will cost to water your garden and power the technology you're putting into your homestead. This goes back to practicing patience.

Without help from the BC Greenhouses team, it would have taken me weeks to get the greenhouse up and running.

The first grow bag garden at the homestead allowed me to see how things grew at the new property with room to move and adapt for the next season.

I put in a garden my first year at the Epic Homestead because I think the garden carries most of the value of the homestead. I didn't put the planning in that the project really needed. I knew I was rushing into things, and I put in what I consider "lightweight" versions of garden beds: small in-ground beds and raised beds that I could disassemble—nothing permanent. I used that first year as a learning year, and I changed up the design of the gardens at least three times before I got the bed placement and irrigation setup just right. I don't regret this, because I wanted to start growing food right away, but I see how I could have saved myself time and energy if I'd waited.

Pick and choose the projects where you're going to invest your time and money, and slowly, you will build a homestead that you're really proud of.

DESIGNING YOUR HOMESTEAD

In a lot of ways, designing a small, urban, or suburban homestead is more difficult than designing a more sprawling rural place. Urban and suburban homesteaders have to work within their zoning regulations, worry about close neighbors, and use their space as efficiently as possible. Efficiency and smart design are important for rural homesteaders, too, but there's more room for error on a larger property.

EPIC TIP

STAY ORGANIZED

Staying organized with all that's going on around a homestead is a task in itself. I use Notion, an app that's free for personal use. It's robust and flexible, so I can make it anything I want: a project manager, a to-do list, a notetaking app, a journal. Any task management system, even if it's just pen and paper, can work for you.

Sometimes you're just going to have to design a homestead in the only way that your constraints will allow, even if it's not the ideal way. The following is my advice for working out how to situate your homestead.

Map It Out

It's great to have a to-scale map of your property. Unless you have a landscape-designer friend who wants to draw out your property using their mapping software, all you need is a big sheet of graph paper. This map doesn't need to be perfectly to scale, but the grid boxes will help you see the scope of your homestead.

Have fun with this: Play around with different layouts, get your family into it, draw your sun path and the areas of shade on it, and map out irrigation lines. If you were to draw a 3-by-4-foot (0.9 x 1.2 m) raised bed on your map to scale, you would be able to see how it interacts with other areas of your property. This overview allows you to think more strategically about the projects you're considering. You can get really detailed with this map to see what makes the most sense for the space you're working with before you even put a shovel in the ground.

And before you put a shovel in the ground, sit with your map and your plan. When I was into art, I'd draw for hours and think it was amazing. Then I'd wake up the next day and think it was total garbage. This self-criticism is part of being an artist, yes, but it's also what's involved in creating anything new. A change of heart can happen in your homestead planning phase, too—except here, the results are more permanent than those sketches I used to make!

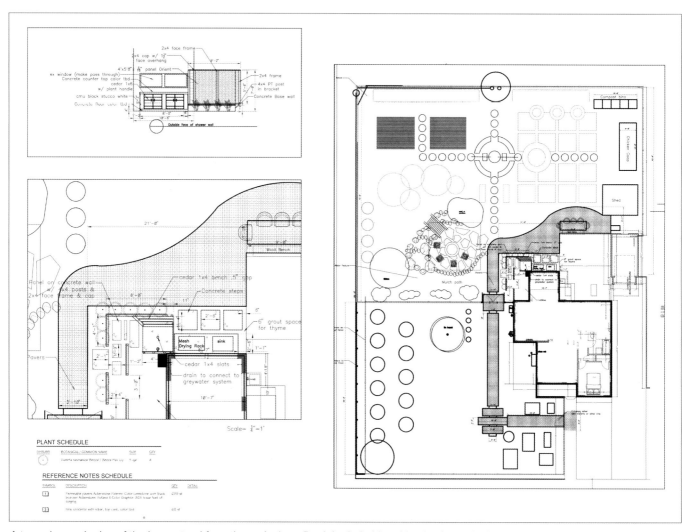

A to-scale rough plan of the homestead from the early days. Don't be intimidated by the fancy design—sketching on graph paper will do.

Almost as if by fate, the only mature tree already planted at my new homestead was a loquat tree, which was the only tree at my prior property.

Use Your Resources

The property you're homesteading on may already have structures and plants in place: a house, perennial plants and trees, water lines, fencing, a garage, and other buildings. You can look at these as a help or a hindrance. Unless you're planning on tearing it all down and starting with a clean slate, I suggest you use all these to your advantage. Create microclimates, and use the layout as a jumping-off point for designing your homestead.

The sunlight available to your property is a very important resource and a huge factor in where to place important elements of your homestead. Your garden and orchard need adequate sunlight to grow. Your solar panels need adequate sunlight to capture energy. Your chickens, water reservoirs, and compost rely less on direct sun exposure.

The simplest way to figure out how much sun your yard gets is observation: Go outside in the morning, afternoon, and evening, and look at the yard to see where the sun pattern falls. As the Earth moves around the sun throughout the year, the light patterns will change. The website Suncalc.org can help give you a sense of the length of daylight and direction the sun is coming from throughout the year. Type in your address, and choose the date you'd like to see. Scroll through the hours of the day to watch the sun move across your property. This website doesn't take into account trees and structures, so you'll need to use your imagination to get a good picture of your sunlight availability.

Add the expected shade to your property map to better visualize how existing and planned structures and trees will change the availability of sunlight. This is important. I didn't actually do this when I put in the new garden shed, and I just got lucky that it didn't throw a shadow over everything. You don't have to rely on luck.

Resources go beyond what's on your property and extend into your community. Think about what your area is rich in. California, specifically San Diego, is rich in sunlight. I wanted to take advantage of this strength by installing solar panels to use tax credits, wipe my energy bill to zero, and recoup my costs in just a few years. Your community resources might include rebates for climate-smart projects, abundant rainfall, or a community tool library stocked with garden and farm tools.

Design from the Outside in

Maximize the use of the space you have by pushing your homestead to the boundaries. For apartment homesteaders, this is using your windowsills and balcony railings. For anyone with a small lot, it means butting up against your property line whenever possible.

The shed on the property when I moved in was too small and awkwardly located.

Building the new garden shed close to the house and property line was an obvious design choice that made the rest of the garden easier to plan.

I tend and harvest from the raised-bed veggie garden daily, so building it close to the front of the house made perfect sense.

When I moved to the Epic Homestead, the existing garden shed was placed a good way off the fence. This didn't make sense to me, because the space between the fence and the shed was awkwardly shaded and not really useful. I put up the new shed just 5 feet (1.5 m) off the fence—enough space to walk behind and have some hidden storage but not so much space that anything was wasted.

Use your first installation to judge the placement of your next project. With the shed in place, I saw a better layout for my garden and knew where the chicken coop and compost bin fit in. Both the coop and the compost bin are close to the property edges, too. By starting at the edges and working your way in, you'll use your space more efficiently.

Keep Busy Areas Close

The more you plan to engage with a part of your homestead, the closer it should be to your house. This is a central idea in permaculture design. Permaculturalists look at their land in zones—concentric circles radiating away from the center of activity (your house)—and put their most-trafficked areas in the inner zones, working outward. Without being attached to permaculture concepts here, this idea makes sense to me. Look at this from the point of view of growing food:

- You'll want herbs from your kitchen garden nearly every day, so those should be close to your kitchen. If you have to walk through the chicken pen and around the compost bin to get to the herbs you want, you would be tempted to skip it and go without fresh herbs.
- Harvesting and working in your vegetable garden happens a few times a week, so that might be the next-closest food-growing area.
- Your orchard needs little attention once it's established, so that could be farthest away.

Longer-season crops, like garlic, are grown in-ground and further away from the house, as they need less daily attention.

Where to Put Your Garden

Because of the essential nature of the garden, its design, layout, and location are very important. A well-placed garden could mean the difference between a great harvest and an OK harvest.

One thing every garden needs is sun, and plenty of it. It's easier to construct a shade structure than it is to manufacture sunlight. If you're in the northern hemisphere, orient your garden so it gets southern exposure. Most of the vegetables you want to grow prefer 6 to 8 hours of direct sun each day.

If you have to build your garden on the north side of a structure or behind a tall wall, you might be working with a shade garden and have to choose your plants accordingly. This garden can still produce leafy greens, some herbs, and cooler-weather crops, but it won't give you epic tomatoes, peppers, and other classic summer vegetables.

In looking at the parts of your homestead where you'll spend a lot of time, the garden is at the top of the list. Keep it close to your house so it's not a burden to get there. Of course, if you have competing interests—maybe the area closest to your house is actually in the shade—choose the options that make the most sense for the food you're hoping to produce.

Study the placement of the structures closely and keep in mind that the house is to the right of this image. What do you notice?

The garden goes hand-in-hand with other homestead systems, as well. Keeping the compost bins close by will simplify garden cleanup. Having chickens next door makes it easy to dispose of tasty scraps in their enclosure. And don't forget about irrigation: Look at how far you have to run hoses and irrigation tubes as you make placement decisions.

Whether in California or in Norway, a homestead can be made to work anywhere—especially now, with the technology advances that we have to create comfort in modern homesteading. You might look at ten houses, like I did, before you find your spot, or it might be more like twenty. Wherever you end up, you can learn your climate, get to planning, and make your homestead epic.

The compost system is near the chicken coop for compost ingredient access and easy cleanout.

Nothing on this Earth is more satisfying than harvesting produce you've grown yourself for friends and family.

chapter two

OUTDOOR FOOD GROWING

MY FIRST GARDEN was hard to compare to anything I'm doing now. It was part-hydroponic, part-container garden, and I had no knowledge of soil health or plant biology. I just went to the store, bought whatever plants looked cool, put them in pots, and hoped they grew. This is how gardeners often start out, which is totally normal!

I have now learned from my mistakes and will be more intentional about the design of my systems. That's what I'm trying to do here at the Epic Homestead, and that's what this chapter—and this whole book—is about.

There is no one-size-fits-all garden. Consider the purpose of your garden: Are you trying to grow 100 percent of the food for your family, or are you trying to get a nice flower garden or a nice kitchen garden for your homestead?

Starting out, don't overwhelm yourself. Give yourself a manageable garden, and build on it each season as you figure out your plan. I would not start with more than two raised beds per person.

THE FOUNDATION OF THE HOMESTEAD

A lot of homesteading focuses on large projects that you need to build once and pay little attention to after that. The garden, though, is a never-ending project. There's so much you can do there every day of every week. Every season, you'll continue learning and trying new crops and growing techniques. The garden is a place of growth.

I spend at least 15 minutes a day in the garden. This isn't always hard work; sometimes I'm just getting herbs for cooking and having a look around. If I'm planting out seedlings or turning over beds, that's an hour or two of my time. On average, I probably spend 5 to 10 hours a week in the garden. For the food and enjoyment I get from the space, that's not a lot of time.

Scan for video

My old front-yard urban garden. This is proof that you can grow in small spaces.

In return for consistent time and attention, this area of the homestead provides in multiple ways. From a tangible self-sufficiency angle, the garden feeds us with vegetables and herbs to eat fresh and preserve for later. This produce is fresher and more nutritious than anything you can find in a grocery store, and you know exactly how it was grown. This is the most obvious value of the homestead garden.

Looking at this using monetary value, consider what it would cost for you to buy all the food that you'll harvest. Add in the cost of foods you preserve, like canned tomatoes and dried herbs. It's likely that no single other area of your homestead offers as great a monetary return as your garden. The exception here might be if you have a large, mature orchard that is producing a lot of fruit, because fruit sells for a high price.

The garden acts as a place of being that allows you to decompress from some of the more stressful and stimulating parts of modern society. While I live this high-tech, natural lifestyle and encourage you to use technology to enhance your homesteading experience, too, the garden is a place to unplug from the pace of modern life that we're sucked into.

Don't underestimate the psychological and social aspects of gardening. Here, you can tie in community and family time. You'll learn a lot about yourself and deepen your relationships with others. Look to the garden to teach the children in your life math, science, biology, and practical skills.

Then there's the physical element of gardening. This time outdoors is beneficial, providing you with exercise and sunlight, which offers essential vitamin D.

Your chickens, bees, orchard, and other areas are all part of the tapestry of your homestead, but your garden acts as the foundation. Gardening is important enough to me that this is how I began my homesteading journey, with Epic Gardening on the blog, YouTube, Instagram, and more. Most other areas of homesteading covered in this book interact with the garden in some way.

Tending to seedlings in the greenhouse is an exercise in plant care and mental relaxation.

GROWING METHODS

A beautiful thing about plants is that they want to grow. As you've seen with spindly grasses on a windy sand dune and dandelions coming up through cracks in a sidewalk, if conditions are at all right for a seed to germinate, it will. You have many options for cultivating a garden that are kinder to plants than having them grow in pavement and salty sand. One or more of the growing methods covered here can work for your homestead.

Raised Beds

Aside from simple container gardening, raised beds are probably the easiest way to get started gardening. Raised beds are not permanent, which is ideal for renters. These structures can be as fancy or as basic as you'd like.

Raised beds are my preferred method to grow food in a simple, repeatable, and scalable way.

At the edges of my raised bed garden, I include pollinator-attracting plants like sweet alyssum, sunflowers, and more.

Thirty inch (76 cm) tall raised beds make gardening a breeze for someone like me (a 6'4" [193 cm] guy who doesn't want to bend over).

Raised bed garden design. Anyone who's followed my work knows my love of Birdies Garden Products' coated-metal raised beds. They're the original metal raised bed and solve a lot of problems:

- They won't rot like even high-quality wooden beds.
- The tall design makes them easy to work from.
- They are easy to build and move around, as they're quite lightweight.

Preconstructed beds like these make raised-bed gardening simple.

For DIY homesteaders, building your own can also be pretty easy. In my first book, *Field Guide to Urban Gardening*, I covered three build-your-own raised bed plans. I don't want to rehash all of that here, but I will say this: You can build a raised bed in 20 minutes for less than $50 USD using all new materials. All you need to get started with a 2-foot-by-6-foot (61 x 183 cm) raised bed are four planter wall blocks from a hardware store, 6 cubic feet (170 L) of soil mix, and four 2-by-6-inch boards: No nails, no glue, no construction experience required.

Filling tall beds up to 50 percent with garden debris saves you a ton of money on soil.

Whether building or buying a raised bed, go for one that's appropriate for your physical ability. A raised bed that's 6 feet (1.8 m) across will require a lot of reaching to do any work in the center of the bed and is probably too wide, even for someone as tall as me. A better bet would be 4 feet (1.2 m) across, or even 3 feet (1 m).

Consider the same for the bed's height. I prefer 15 inches (38 cm) as my minimum raised bed height. Taller raised beds are great for more ergonomic gardening, for gardening that's accessible to gardeners with limited mobility, and for growing plants with deep tap roots—including daikon radishes and extra-long carrots. Plant roots need room to grow, so you don't want to go shorter than 6 inches (15 cm). You also need to fill the bed with a growing medium (more to come about this), and the cost of this will increase with the depth of the bed.

If you're working with a small area, you may be tempted to maximize the space that you can grow in by cramming raised beds together. I'm here to tell you this isn't the move. I did this in my old space, and it was cramped! It worked, but it was annoying. Without enough room to maneuver between raised beds, you'll need to bend and lift in unnatural ways—it's a quick way to injure yourself. Butting beds close to one another can also cause tall plants to shade out neighboring beds. Leave about 2 feet (61 cm) between raised beds so you have plenty of space to work and move between them.

Filling raised beds. You have a few options for filling your raised bed with a growing mix. Really invest in your raised bed growing mix, because the health and productivity of your plants relies on the health of your soil. You'll read more about that on page 44.

You can purchase ready-made raised-bed mix, or you can make your own raised-bed mix using a recipe from the Epic Tip: Make Your Own Raised Bed Mix (see page 34).

For the budget-conscious, making your own soil mix is dramatically cheaper than filling with bagged mixes.

EPIC TIP

MAKE YOUR OWN RAISED-BED MIX

Go DIY with a growing medium for your raised beds using one of these recipes:

Mel's Mix™

This raised-bed soil recipe comes from Mel Bartholomew, who developed the Square Foot Gardening Method™ (and, incidentally, was my garden mentor).

- ⅓ blended compost
- ⅓ aeration component (like perlite or vermiculite)
- ⅓ water-retentive material (like sustainably sourced peat moss or coconut coir)

Native Soil Mix

Joe Lamp'l, another gardener I admire, suggests this mix of materials for raised beds.

- ½ topsoil
- ¼ compost
- ¼ grass clippings or unfinished compost

Raised beds aren't meant for tilling. Add nutrients by top dressing the bed with compost, almost like a mulch, or use a rake and mix the compost into the top 2 to 3 inches (5 to 8 cm) of soil. As you water, or the bed is rained upon, the nutrients will filter through the soil and into the root zone.

If you have super-tall raised beds, the cost of growing mix adds up. Use the beds' height to your advantage by filling it partway with organic matter and topping it off with the raised-bed mix. This is a combination of hügelkultur (read more about that in Alternative Growing Methods, page 43) and advice from my friend Mark Valencia, of Self Sufficient Me.

Most annual plants have root systems that only reach 12 to 18 inches (30 to 46 cm) into the soil. Measure to that depth from the top of the raised bed, and fill the space below with organic material. You could use vegetable scraps, horse manure, cow manure, material from a passive compost pile, wood bark, sticks and twigs, grass clippings—you get the idea. Larger-volume pieces of organic matter, like logs, are nice because they take up even more space, and they're slower to break down. Just add more compost to the top of the bed as the organic materials break down and the level of the raised-bed medium gets lower over time.

Whichever method you use to fill your raised beds, leave a couple inches at the top for mulch. Mulch your raised beds just as you would in-ground beds to protect your soil from too much sun and to retain moisture. Weed pressure is less in raised beds, but mulch will help to keep weeds away, too.

Keep your soil covered, even when transitioning between seasons. Add a layer of compost or mulch over the winter, or grow a cover crop that you can use as a mulch or incorporate into the soil next year.

Containers

For homesteaders in apartments and with small yards, containers are the way to go for growing food and flowers. Even for me, with more space than I've ever had, I still grow some things in containers.

Container gardens are great for getting perennial plants started before you know where you want to plant them for good. They're excellent for adding life to a porch or patio. For gardeners in colder climates, growing specialty crops in containers means you can bring them indoors when frost is on its way, which could mean the difference between getting a harvest and not getting a harvest. Using containers in this way also allows northern gardeners to grow herbs like lemongrass and bay laurel as perennials instead of annuals.

Container choices. While container growing is easy, selecting the right containers to grow isn't as easy. There are so many options now, including self-watering pots and containers with trellises built in.

It's no secret that I'm a fan of grow bags—I wrote a whole book about them, *Grow Bag Gardening*. Grow bags are lightweight fabric pots that come in a variety of sizes from several different brands. Unlike solid-sided containers, grow bags encourage plants to develop a stronger root structure. They usually have handles, so you can move them around as needed. Because they're made of fabric, you can fold them up and store them easily between seasons. Grow bags also regulate soil temperature and water retention better than pots made of other materials. On the downside,

Lined grow bags help retain a bit more moisture than unlined bags but retain all other benefits.

Screen old potting soil through ¼ inch (6 mm) chicken wire.

Amend with worm castings, fresh potting mix, or other fertilizers.

Add revitalized mix back to the containers and get growing!

grow bags are generally more expensive and need to be replaced more often than other garden containers. Long, rectangular planters are great for homesteaders who have small areas. For those living above others—in the case of balcony gardening—do your neighbors a favor and have some kind of saucer to collect drippings from the containers. Balconies and porches offer space for planters designed to sit on top of the railing. Under porch eaves is space for hanging planters, if that area gets enough sun for what you want to grow there.

Filling containers. There's this myth that you need to put drainage material at the bottom of your containers. Soil with gravel underneath actually retains more moisture than soil with finer particles, like sand, mixed throughout. Instead, use high-quality potting soil, water your plants well, and be sure your pot has a drainage hole.

You'll find a lot of mixes marketed as container "potting" soil. Here's what I look for when choosing a growing mix:
- Lighter, finer-textured mixes are best for use when starting seeds and rooting cuttings.
- Mixes containing a high percentage of coarse sand or pine bark are best for potted trees and shrubs.
- DIY potting soil with a sandy or gravelly texture is ideal for cactus and succulent growing.
- When growing a mixture of annuals, perennials, vegetables, and tropicals, the best fit is a general, all-purpose potting mix.

It's not necessary to refill your containers with fresh soil mix every season. Rejuvenate your spent soil to improve soil quality, texture, and nutrients as needed. There are two ways to do this, depending on the condition of the soil.

Here's a simple way to do a refresh every year:
- Dump your spent soil into a larger container.
- Break up clumps of dirt and remove any plant debris.
- Add fresh potting soil or compost—gardener's choice—that equals about one-third the volume of the container. So, if you have a 5-gallon (19 L) pot, you'll add a little more than 1½ gallons (6 L) of fresh material to the large container.
- Mix all of this together, and moisten the mix.
- Put it back into your growing container.

If your soil is tired—maybe you grew tomatoes in the container or didn't get around to refreshing it last season—try an epic refresh:

- Remove all the plants and plant roots. (In Soil Building, page 44, I encourage you to leave the roots in place. That's for raised beds and in-ground beds. There's just not enough room in containers to accommodate all of past seasons' roots, and they won't break down as quickly in containers.)
- Run the soil over a soil sifter to break up the chunks. Look for grubs and other pests that are trying to overwinter in the soil and remove them.
- Add fresh potting soil or compost that equals about one-third the volume of the container, as above.
- Add slow-release fertilizer to keep the soil nutrients balanced over time. Chicken manure, blood meal, and bone meal are good choices.
- Add worm castings. If you're making your own vermicompost, you can harvest right from your bin to add microbial life and all-purpose fertilizer to your containers.
- Mix and moisten.
- Repot your containers.

Watering containers. Overwatering is the biggest issue in container gardens. Especially if a container doesn't have a drainage hole, it's easy to oversaturate the soil and leave plant roots to sit in too much water.

How much and how often you need to water depends on the temperature, sun exposure, and air movement. The warmer, sunnier, and breezier, the more water will both evaporate from the soil and be taken up by the plant roots. Smaller containers dry out faster than larger containers, and porous containers—terra cotta pots and fabric grow bags—dry out faster than impermeable containers, like plastic pots.

EPIC TIP

DIY SUB-IRRIGATED PLANTER

Cut up natural sponges and layer them across one-third of the bottom of a container. This will hold moisture and then release it back into the soil as the soil dries out.

You can, of course, purchase ready-made sponge-like inserts made to fit your container, as well.

In-Ground Beds

Now that I have a larger space at the Epic Homestead, I've cultivated a love of in-ground gardening. For homesteaders with a plot of land, in-ground beds are the cheapest gardens to start. You don't even need to build or buy anything to create one.

The biggest downside to an in-ground garden is that it requires a lot of bending and lifting at ground level. I like to counter this by growing taller plants in the in-ground beds so I can work with them at my standing height.

There are just three stages to getting started with an in-ground garden: soil assessment, bed building, and planting.

Start with a soil test. Soil tests are important because before you know what's needed to improve your soil, you need to know your baseline levels.

A simple at-home test can check the nitrogen levels in your soil.

You can have your soil tested for heavy metals, macronutrients, micronutrients, organic matter, pH, and more. To start off a home in-ground garden, the more specific you are in your testing, the better you'll understand what amendments your soil needs to grow great plants. If you are growing in urban areas and areas that have questionable uses, be sure a heavy metal test is on your list for food safety.

Contact your County Extension office for a basic soil test, or send a soil sample to an independent soil-testing lab for more complete tests. Some labs offer specific amendment recommendations to balance your soil. We'll get into amendments later in this chapter.

Till, or don't till. Tilling is the act of breaking up the soil surface. You can use a rototiller or the old-school double-digging method. Tilling is helpful for breaking up compacted soil, burying some weeds, and incorporating plant matter that will break down and feed the soil.

But I've been inspired by no-till gardeners like British gardener Charles Dowding and now I want to disturb the soil as little as possible.

While tilling can reduce compaction in the top layer of your soil, it can also increase soil compaction of the layer below. If you till the top 6 inches (15 cm) of your garden, at a depth of 7 inches (18 cm) and below, things can actually get worse. This hardpan can stunt the growth of deep-rooted vegetables.

Tilling eliminates the weeds existing above ground, but it also stirs up the weed-seed bank, bringing new weed seeds to the surface. It also chops up the roots of weeds that propagate via rhizomes. In my case, that's Bermudagrass, and tilling caused a Bermudagrass nightmare in the front yard of the Epic Homestead. Think about it: If your tiller chops a 5-foot (1.5 m) long piece of rhizome into ten different pieces and mixes it into your soil, you just planted ten more weeds.

One last downside of tilling is that each time you mix up the soil in this way, you shred the mycorrhizal fungal networks, the earthworm tunnels, and the invertebrate life.

Having weighed the pros and cons of tilling garden soil, I like to till once to get the garden bed established and incorporate amendments, and then go no-till.

Collecting a representative soil sample and mixing it together.

Mixing the sample with water prior to testing.

HOW TO BUILD A BASIC IN-GROUND GARDEN BED

Measure out an in-ground bed width. This one is about 30 inches (76 cm).

Materials
Twine
Landscape flags
Measuring tape
Rototiller
Soil amendments
Raised-bed soil mix
Cardboard
Woodchips
Seeds or seedlings

To start an in-ground bed in a new area, you have the option of tilling the soil and planting into it right away or covering it with cardboard and mulch and waiting a season before planting. This project is a step-by-step guide of how I build in-ground beds by tilling.

Building Instructions

1. Measure and mark the perimeter of your full garden area—pathways and all. Plan to till everything now and build pathways in a minute. For small-space gardeners, you may only be marking the space for a single in-ground bed.

2. Till 4 to 8 inches (10 to 20 cm) of the top layer of soil across the whole area. You may need to run the rototiller across the area more than once to break up very compacted soil.

Bust out the tiller—my one-time recommendation for hard, dead soil.

3. Measure and mark pathways and beds. Stakes and twine or landscaping flags and string lines work great here. Create pathways between beds that are wide enough to get a wheelbarrow through—2 to 3 feet (61 to 91 cm) is a good width.

4. Dig out the soil from the pathways, and add it to the garden beds. This raises the level of the garden bed soil, which allows for better water filtration and more depth for plant roots to grow. Once you till, keep foot traffic off the garden beds to keep from re-compacting the soil.

5. Apply amendments and organic matter to the garden beds.

6. Lightly till the garden beds again to mix in the amendments. Don't till the pathways.

7. Top the garden beds with 1 to 2 inches (2.5 to 5 cm) of high-quality raised-bed mix.

8. Smooth out the bed surface with a rake.

Add fresh compost to the top of the freshly tilled bed.

Rake in compost, worm castings, and fertilizers.

Generously water the newly-shaped bed.

Plant the bed. I used classic Thai cooking ingredients.

9. Water the bed well, and let it rest for a week or so. This downtime allows nutrients to filter into the root zone and gives soil life time to populate before you put plants in the ground. Pull any weed seedlings that start popping up during this time.

10. Improve your pathways. Cardboard covered with wood chips is a great option to reduce weeds and maintenance in pathways. As the wood chips break down, the nutrients feed the soil around the pathways. In areas with more abundant water, you can plant living pathways with something like white clover or chamomile—low-growing plants that withstand being walked on—and plan to mow them for control.

11. Plant your beds. This is the fun part. Measure rows and plant according to seed packet instructions.

EPIC TIP

RENT A TILLER

Larger garden equipment, like a tiller, is expensive. If you're only using the equipment once per season—or once for the life of your garden—save money and storage space, and rent the equipment. Look to equipment-rental companies, big-box hardware stores, and your local university or governmental agricultural office for rental equipment.

ALTERNATIVE GROWING METHODS

By no means does this book cover all the ways that you can grow food outdoors. There are methods from around the world that use ancient knowledge and scientific thinking that I don't even know about. A few growing methods in addition to the ones in this chapter include:

HÜGELKULTUR

The name for this German growing method translates to English as "hill mound." Hügelkultur beds are first dug 6 to 12 inches (15 to 30 cm) deep. This pit is filled with large pieces of rotting wood, then the sod that was removed, layers of coarse compost and other organic matter, and fine compost and soil on top. All of this is tamped down and planted into. These beds stand 3 feet (91 cm) tall and are 3-D, making even more space for planting. As the organic materials in the mound decompose, the hill gradually settles, and the interior holds moisture for the plant roots.

KEYHOLE

Keyhole gardening was developed in Africa to use active composting to retain moisture and supply nutrients to the raised bed. Instead of digging a hole and creating a mound for compost, as in hügelkultur, a keyhole garden is built in a circular or square shape around the compost pit. A notch is built into the garden—that's the keyhole— so the gardener can access and easily add to the compost pile. The "U" shape of the garden bed allows you to get a lot of square footage while being able to reach all of the planting space.

HYDROPONICS

Often thought of as an indoor food-growing technique, hydroponics can be done outdoors, as well. Hydroponic growing definitely fits into the "high-tech, natural" homesteading category. Outdoor hydroponics has the right balance of nutrients and water without needing the climate control and artificial lighting of indoor setups. Plants in outdoor hydroponics situations can grow faster and healthier, and you won't need to weed them because they're growing in water, not soil. You can build your own hydroponics system with 5-gallon (19 L) buckets or go epic with fully outfitted kits.

FOOD FOREST

A food forest is essentially the opposite of what industrial agriculture looks like. You might also know it by the name polyculture or forest garden. A food forest is a diverse planting that more closely mimics plant ecosystems we see in nature. You'll read more about food forests on page 88.

A vertical *nutrient film technique* hydroponic system for lettuce.

Here, I'm using a machete to chop and drop the backyard cover crops with Jacques.

SOIL BUILDING

In a healthy garden, we don't feed our plants. We feed our soil and the life within it, and that's what feeds our plants. This is the number-one tactic for building a happy, healthy garden, whether that's an in-ground garden, container garden, or raised-bed garden. I've come to appreciate using a mix of different soil-building techniques at the Epic Homestead.

Grow Cover Crops

A lot of homesteaders don't consider cover crops, but they're a mainstay of market gardeners, and they're becoming more popular with commercial-scale farmers, too. I think of cover crops as the laziest way to improve your soil. Growing a "green manure" keeps your soil covered when you're not growing a regular crop—preventing erosion and nutrient loss—and keeps roots in the soil to feed the soil life. When you cut down the cover crop, that debris breaks down and puts organic matter back into the soil. Be sure to cut down the cover crop before it goes to seed, otherwise you may be growing another season of the cover crop, whether you intended to or not.

The Epic Chickies provide not only eggs, but also nutrient-rich droppings that I use in the garden.

Use a mix of cover crop types that makes the most sense for your garden. Types of cover crops include:

Dynamic accumulators like comfrey, daikon radish, dandelion, and others have long taproots that mine nutrients from deep into the soil. They bring those nutrients into their plant tissue and release those nutrients into the top layer of soil as they decompose. The tap roots from dynamic accumulator plants help to break up heavy, compacted soils, and provide aeration, as well.

Legumes are nitrogen fixers. Their roots form symbiotic relationships with nitrogen-fixing soil bacteria. Nodules on the plant roots allow the bacteria to steal nitrogen from the air and put it into the soil in a usable form. Clovers, peas, beans, alfalfa, and hairy vetch are in the legume category.

Brassicas in flower attract beneficial insects, which go after pest insects in the garden. Brassica species also release chemical compounds toxic to pest nematodes, fungi, and even some weeds. Mustard, collards, radishes, and turnips are brassica cover crops. These are fast growing in cooler weather, making them great spring and fall cover crops.

Grains and grasses, like wheat and rye, can grow tall, providing a lot of organic matter and a great mulch after being cut down.

Get Chickens

Chickens make use of your garden and food scraps and produce a valuable compost ingredient. During the off season for your garden, let your chickens into the growing space. Their droppings are full of beneficial bacteria and nutrients. Their natural scratching habit works the chicken waste and the food scrap waste into the top layer of soil, too.

Save the Roots

When you're turning over a garden bed—whether that's a raised bed or an in-ground bed—chop down the previous crop at the soil level and leave the roots in the soil to give the soil life more to work with. Roots left behind decay and are consumed by earthworms, beetles, bacteria, and fungi. The air pockets left by the roots allow soil aeration and break up compacted soil.

Amend the Soil

When you harvest vegetables or remove vegetative material from a garden bed, you're removing with it all the nutrients that the crop absorbed during its growth. Now it's time to add that back with soil amendments. If you're looking for an all-around balanced soil booster, homemade compost is the least expensive option. There's a whole chapter on composting in this book, see page 107.

Depending on what your soil test reveals about your soil deficiencies and on your own observation of how your plants are doing, you may need an amendment that provides a more specific solution.

A few amendments to know include:

Feather meal is a slow-release, high-nitrogen fertilizer sourced from ground-up poultry feathers.

Rock phosphate is a mined rock that makes a good source of slow-release phosphorus.

Cow manure is a balanced source of nutrients and can be added to compost or applied to the garden on its own. Because animal waste contains pathogens, follow the industry-standard food-safety rule known as the 90/120 rule. I cover this on page 129.

Chicken manure is probably easy to find, whether from your own chicken coop or a neighbor's. Calcium, magnesium, sulfur, and micronutrients, like copper and zinc, are plentiful with this amendment. Chicken manure is a great addition to a compost pile, too.

Lay Mulch

Mulching has so many benefits to the garden: weed suppression, water retention, and soil heating and cooling. There's one more longer-term benefit, which is that organic mulches break down over time and release their nutrients into the soil. The wood chips in pathways and straw mulch in garden beds all eventually return to the earth as food for soil life. This is a slow but important way to improve your soil.

Massive piles of arborist woodchips in the driveway were a common occurrence in the first year of building the homestead.

Turning the neighbor's trash into my treasure. These chips were dropped in my yard for $0.

EPIC TIP

USE FREE WOOD CHIPS

In Chapter 1, I wrote about using the resources already available on your homestead. This applies to resources available in your community, as well. In most places, tree-trimming services are hard at work cutting down and mulching trees. This is a waste product for them and a free resource for you. A few ways to find free wood chips:

- Flag down a tree-trimming truck working in your neighborhood.
- Call an arborist in your area.
- See if your municipality offers free wood chips.
- Use ChipDrop.com to find free wood chip mulch.

Over 40 yards (36.5 m) worth of organic matter to improve the dead soil on the property.

A mix of annuals and perennials balance the garden harvests.

WHAT TO GROW

I am often asked how a gardener should decide what to grow. My best advice is to pick the things that you know you like to eat. Especially if you're growing in a small space, dedicate your garden area to the crops that you'll find enjoyable.

If you're just getting started or don't have a lot of time to dedicate to your garden, pick the easiest crops to grow. Potatoes and beans are two of my favorite low-maintenance crops.

The Garden Production Chart (on page 49) can help you determine how much of each crop to plant. The thing about a chart like this is it's literally impossible for me—or for anyone—to tell you how much of any crop you'll be able to harvest. You might have a great year for kale with cool weather and just enough rain, but that same year could be too cloudy and damp for tomatoes. Every year is different, so use the chart as a guide, but be flexible in your planning.

If a homesteading goal is to preserve food for your family to eat year-round, pay attention to the crops that are good for canning and freezing. There's more about all of this in Chapter 9.

Annual Crops

Annual crops are those that have a one-year lifecycle. Whether a crop is an annual or a perennial depends in part on your climate. Tomatoes, for instance, are grown as annual plants in most areas, but Jacques Lyakov, who you read about on page 11, managed to overwinter a huge tomato plant that continued producing the next year here in San Diego. Most vegetables that we grow in home gardens are grown as annual crops.

Perennial Crops

Perennial crops are those that you plant once and harvest from year after year, including trees, shrubs, and a lot of fruits, herbs, and flowers. You can grow many of these in pots indoors and on balconies just as well as you can in an in-ground or raised-bed garden. Because you plant these once and don't have to disturb the garden bed again, keep perennials separate from annuals in your planting areas.

HOW TO USE THIS CHART

The pounds listed in the chart at right are what an adult who really likes this vegetable might eat in a year. If you think eggplant is just okay, you'll probably need to plant less. If you really, really love peppers, you might want to plant more. Multiply these numbers by the number of adults your homestead is feeding. Multiply by 0.5 for kids.

GARDEN PRODUCTION CHART

VEGETABLE	AMOUNT NEEDED FOR ONE ADULT (FRESH USE)		AMOUNT NEEDED FOR ONE ADULT (PROCESSED/STORAGE)	
	Pounds (kg)	Sq. Feet (m²)	Pounds (kg)	Sq. Feet (m²)
Asparagus	1.5 (0.7 kg)	30 (2.8 m²)	5 (2.3 kg)	100 (9.3 m²)
Beans, snap	15 (6.8 kg)	44 (4.1 m²)	18 (8.2 kg)	53 (4.9 m²)
Beets	3.5 (1.6 kg)	6.25 (0.6 m²)	7.5 (3.4 kg)	13.25 (1.2 m²)
Broccoli	8 (3.6 kg)	50 (4.6 m²)	12 (5.4 kg)	75 (7.0 m²)
Brussels sprouts	6 (2.7 kg)	50 (4.6 m²)	8 (3.6 kg)	66.75 (6.2 m²)
Cabbage	15 (6.8 kg)	31.25 (2.9 m²)	15 (6.8 kg)	31.25 (2.9 m²)
Carrots	10 (4.5 kg)	14.5 (1.3 m²)	10 (4.5 kg)	14.5 (1.3 m²)
Cauliflower	9 (4.1 kg)	27.25 (2.5 m²)	12 (5.4 kg)	36.25 (3.4 m²)
Celeriac	0.5 (0.2 kg)	1.5 (0.1 m²)	—	—
Celery	4 (1.8 kg)	2.25 (0.2 m²)	—	—
Chinese (Napa) cabbage	2 (0.9 kg)	1 (0.1 m²)	—	—
Corn, popcorn	—	—	4 (1.8 kg)	40 (3.7 m²)
Corn, sweet	25 ears	69.5 (6.5 m²)	50 ears	139 (12.9 m²)
Collards	2 (0.9 kg)	4.25 (0.4 m²)	4 (1.8 kg)	8.75 (0.8 m²)
Cucumbers	8 (3.6 kg)	16 (1.5 m²)	10 (4.5 kg)	20 (1.9 m²)
Eggplant	4 (1.8 kg)	8.75 (0.8 m²)	—	—
Endive	4 (1.8 kg)	9 (0.8 m²)	—	—
Garlic	1 (0.5 kg)	5 (0.5 m²)	2 (0.9 kg)	10 (0.9 m²)
Jerusalem artichoke	1.5 (0.7 kg)	2.75 (0.3 m²)	1 (0.5 kg)	5.25 (0.5 m²)

VEGETABLE	AMOUNT NEEDED FOR ONE ADULT (FRESH USE)		AMOUNT NEEDED FOR ONE ADULT (PROCESSED/STORAGE)	
	Pounds (kg)	Sq. Feet (m²)	Pounds (kg)	Sq. Feet (m²)
Kale	1 (0.5 kg)	1.75 (0.2 m²)	2 (0.9 kg)	3.5 (0.3 m²)
Kohlrabi	1.5 (0.7 kg)	3.5 (0.3 m²)	—	—
Leeks	1 (0.5 kg)	2.75 (0.3 m²)	1 (0.5 kg)	2.75 (0.3 m²)
Lettuce	6 (2.7 kg)	15 (1.4 m²)	—	—
Muskmelon	10 (4.5 kg)	35.75 (3.3 m²)	2 (0.9 kg)	7.25 (0.7 m²)
Mustard greens	1 (0.5 kg)	3.5 (0.3 m²)	—	—
Okra	3 (1.4 kg)	11 (1.0 m²)	4 (1.8 kg)	14.75 (1.4 m²)
Onions	8 (3.6 kg)	8.75 (0.8 m²)	20 (9.1 kg)	21.75 (2.0 m²)
Parsley	0.25 (0.1 kg)	1 (0.1 m²)	0.5 (0.2 kg)	2 (0.2 m²)
Parsnips	3 (1.4 kg)	10.25 (1.0 m²)	3 (1.4 kg)	10.25 (1.0 m²)
Peas, shelled	4.5 (2.0 kg)	18.75 (1.7 m²)	7.5 (3.4 kg)	31.25 (2.9 m²)
Peas, snap	1 (0.5 kg)	3 (0.3 m²)	1 (0.5 kg)	3 (0.3 m²)
Peppers	3 (1.4 kg)	6.25 (0.6 m²)	3.5 (1.6 kg)	7.25 (0.7 m²)
Potatoes	2 (0.9 kg)	4.25 (0.4 m²)	4 (1.8 kg)	8.75 (0.8 m²)
Pumpkins	10 (4.5 kg)	16.75 (1.6 m²)	8 (3.6 kg)	13.25 (1.2 m²)
Radishes	4 (1.8 kg)	36.25 (3.4 m²)	—	—
Rhubarb	4 (1.8 kg)	16 (1.5 m²)	4 (1.8 kg)	16 (1.5 m²)
Rutabaga	1.5 (0.7 kg)	2.75 (0.3 m²)	2 (0.9 kg)	3.5 (0.3 m²)
Spinach	3 (1.4 kg)	7.5 (0.7 m²)	5 (2.3 kg)	12.5 (1.2 m²)

GARDEN PRODUCTION CHART continued

VEGETABLE	AMOUNT NEEDED FOR ONE ADULT (FRESH USE)		AMOUNT NEEDED FOR ONE ADULT (PROCESSED/STORAGE)	
	Pounds (kg)	Sq. Feet (m²)	Pounds (kg)	Sq. Feet (m²)
Squash, summer/zucchini	10 (4.5 kg)	14.5 (1.3 m²)	3 (1.4 kg)	4.25 (0.4 m²)
Squash, winter	6 (2.7 kg)	13 (1.2 m²)	3 (1.4 kg)	6.5 (0.6 m²)
Sweet potatoes	3 (1.4 kg)	23 (2.1 m²)	4 (1.8 kg)	30.75 (2.9 m²)
Swiss chard	3 (1.4 kg)	6 (0.6 m²)	4.5 (2.0 kg)	9.25 (0.9 m²)
Tomatoes	24 (10.9 kg)	43.75 (4.1 m²)	36 (16.3 kg)	65.5 (6.1 m²)
Turnips	5 (2.3 kg)	8.75 (0.8 m²)	7 (3.2 kg)	12.25 (1.1 m²)
Watermelons	12 (5.4 kg)	70.5 (6.5 m²)	—	—

The numbers in this chart are based on information from
Michigan State University College of Agriculture and Natural Resources.

My in-ground backyard veggie garden operates more like a small farm than a garden.

Lightly packing seed-starting mix into your seedling trays improves germination, as the seedling's roots will adhere better to the soil.

Seed-Starting Materials

12 Epic six-cell seed starting trays (without drainage holes)

1 inch (2.5 cm) deep 1020 propagating tray

Bag of seed-starting mix

As a homesteader, you're trying to maximize the amount of time available to produce food. By starting seeds indoors, you can get a jump-start on things.

Think about it this way: If you plant Mortgage Lifter heirloom tomato seeds in the ground on your last frost date, you'll harvest your first tomato about 85 days later. If you start your tomato plants indoors a month before your last frost date and transplant the seedlings into the garden on your last frost date, you'll have tomatoes about 60 days after your last frost date. More tomatoes and earlier tomatoes are good reasons to go through the effort of starting seeds indoors.

I've tried a number of seed-starting methods over the years, and this is my favorite.

Seeding Instructions

1. Prepare your seed-starting mix.
Keep it simple, and use a reliable, fine-grain, bagged, seed-starting mix. Seeds germinate best in light and airy mixes. Break apart any clumps so the mix is uniform.

2. Fill your seed-starting trays. For homesteaders on a smaller scale, I suggest using six-cell seed-starting trays. I've tried using 120-cell trays, but there was no way I was planting 120 of something, and they were cumbersome. I eventually designed my own Epic Seed Starting Trays with all the features I wanted. Fill your trays to the top, and pack down the seed-starting mix. Give the tray a tap on the table to settle the mix into the cells, and top off the cells again with more mix.

3. Fill your propagating tray. I recommend setting your seed-starting trays in a 1020 propagating tray. This holds 12 6-cell trays, giving you space to start 72 plants. The tray acts as a water reservoir, allowing you to bottom-water your seedlings.

4. Seed your trays. What you do next depends on the size of your seed. If you have a larger seed that needs to be planted at a depth of ½ inch (1.3 cm) or more—such as sunflower, nasturtium, or melons—make a small depression with your fingertip in each cell of the seed-starting tray. If you have a smaller seed that needs to be planted at ¼-inch (6 mm) depth or less—such as lettuce, basil, or arugula—don't bother making the indentation.

6

Misting your trays post-seeding helps moisten the mix without displacing the seeds.

Drop into your hole or drop onto the soil surface two to three seeds. In the end, you'll thin out the seedlings so there's just one per cell, but putting several seeds in each cell nearly guarantees excellent germination. Gently sprinkle more soil on top, or sprinkle vermiculite on top to keep the soil surface moist. Don't bury the seed too deep—this is probably the biggest mistake beginner seed starters make. Pack down the surface of the soil lightly with your fingertips.

5. Label each seed-starting tray. Write the variety and date so you remember what you planted.

6. Water in the seeds. Use the mist setting on your hose nozzle to lightly hydrate the top layer of the seed-starting mix. Move aside one of the seed starting trays, and fill the propagating tray about halfway with water. The holes in the bottom and sides of the seed-starting trays will allow water to wick into the cells and hydrate the whole cell from the bottom. Make sure you water really well to start the germination process.

9

Look for a white, healthy root system peeking out from the bottom of the tray.

Pop each seedling out—root system intact—and transplant into the garden.

7. Place a humidity dome over the propagation tray. Seed-starting trays will dry out if they're exposed to the air. Remove the dome when you see one set of true leaves on the seedlings.

8. Keep the seedlings watered. Bottom watering is ideal to prevent fungal issues and soil disturbance.

9. Pot up or plant out when needed. As your seedlings grow, they may need a larger space before you can safely plant them in the garden. Pot up the seedlings into larger pots with a mix of soil and compost. If the timing is right, transplant directly into the garden.

10. Sterilize your trays between uses. Rinse the empty trays and wash with warm, soapy water in a bucket. Spray the trays lightly with hydrogen peroxide, and let them dry before stacking and storing.

Next, let's talk about a few options for growing food indoors.

EPIC TIP

SEED-ROOM SETUP

When starting seeds indoors, there are a few things you can do to improve your setup.

- Use an oscillating fan to gently blow air over the seedlings. This strengthens the stems to make hardier plants.
- Put the light source as close as possible to the plants without burning the plant—6 to 8 inches (15 to 20 cm), depending on the light. Read all about using grow lights on page 72.
- A simple timer will turn the fan and lights on and off for you. I keep them on for 16 hours, off for 8 hours.
- Most seeds want to germinate in soil temperatures above 60°F (16°C). I go for 70°F (21°C), to be sure. Use a seedling heat mat to keep the soil consistently warm. Find heat mats that specifically fit 1020 propagating trays or any that are safe to use in moist environments.

Microgreens are the perfect indoor crop. They allow you to start growing some of your own food, even if you're in a tiny apartment.

chapter three

INDOOR FOOD GROWING

WHEN I TALK ABOUT being able to set up a homestead anywhere, I mean it. I know that opportunities for gardening while living in an apartment are limited—I started growing food while living in an apartment, after all. If you're lucky enough to have a patio, it may or may not get enough sunlight to support plants. Maybe there's a community garden nearby. Regardless, inside, while you might not have a lot of space, you can set up conditions to maximize food production.

One thing I want to address here is the obvious critique of "That's not a homestead." My perspective has always been that if your eventual goal is some level of self-sufficiency, even apartment gardening can be thought of as preparation and learning for a future homestead. This *does* qualify as homesteading. You have to start somewhere, and if that somewhere is a closet microgreens setup, that's not for anyone else to judge.

In addition to giving small-space homesteaders an option for growing food, indoor food production is something anyone can do any time of year. In the depths of winter, when everything outside is dead or dormant, northern gardeners can still harvest fresh foods using these indoor options. Likewise, in the heat of summer, when sensitive greens are crying for help outdoors in the South, those growing in a controlled indoor environment will be crisp and ready for your salad.

Scan for video

MICROGREENS

It doesn't matter whether you have no outdoor growing space or you have acres upon acres: Microgreens give you the most bang for your buck in nutrition. They're not complicated to grow, and they're ready to eat in just a week (or a few weeks). But here's the real benefit: Some types of microgreens have been found to contain *four to forty times* more nutrients than their fully grown vegetable counterparts.

Beyond your initial investment in seed trays and a light, the cost of growing microgreens is minimal. Seed prices vary, but you can find low-cost seeds, like salad mix, in bulk for $8 USD per pound—and you won't use anywhere near a pound of seed to fill out a tray. Add a couple of dollars for soil, and you're spending far less than you would buying the same amount of salad mix at the store.

Microgreens Seeds

There is no difference between microgreens seeds and other garden seeds. Because you're planting so many seeds to grow a tray of microgreens—way more than you would to grow a regular crop—bulk buying is more price efficient.

Microgreens Soil

You don't need anything too special for a microgreens soil mix. Because microgreens are plants that don't go through a complete growth cycle, having a nutrient-heavy growing mix is less essential here than in growing full-sized crops.

Fine-grained potting medium works great. Avoid large particles in your microgreens soil, because those are hard for small plant roots to reach around. We're looking for dense, even seedling growth throughout the tray, and a fine-grained mix makes it as easy as possible for that to happen. Potting mix with wood products and growing mediums marketed as garden soil generally won't work here. "Soil" is not even necessary in your microgreens growing mix. Hemp and coconut coir mat are both doable. I've tried soil-less media like this and found the microgreens' flavor a bit lacking, so test this out for yourself.

Microgreens Lighting

You can use a few different lighting technologies. I cover indoor grow lights more in the Using Grow Lights section of this chapter (see page 72).

Once you get the hang of them, microgreens can become almost overwhelmingly productive.

Materials

Soil mix

1020 seed tray (draining)

1020 bottom tray (non-draining)

Seeds of your choosing

Planting Instructions

1. Put 2 cups (473 ml) of water in the seed tray. The water will wick into the soil mix in the next step. Watering from the bottom helps to prevent the fungal issues that can crop up when watering from the top.

2. Add soil mix to the seed tray. Fill the tray to within ¼ to ½ inch (0.6 to 1.3 cm) of the top. This will make it easier to harvest without getting your shears dirty.

3. Lightly compress the soil. Place an empty seed tray on top of the one you just filled, and tap on the empty tray to lightly tamp down the soil beneath it. Smooth out any indents in the soil. Compacting the soil just slightly like this gives the seeds a stable soil base to adhere to, which will help germination. Remove the empty seed tray on top.

The basic materials for growing microgreens can be found at any local nursery.

Evenly seeding your trays is key to avoiding mold and ensuring proper growth.

4. Seed your tray. Seed just one type of microgreen per potting soil-filled tray. Sprinkle seeds evenly over the top of the soil more densely than you'd ever imagine seeding in a garden. Work from corner to corner, being sure to get seeds across the whole surface. Avoid large clumps of seed.

5. Mist with water. A light watering from the top helps the seed adhere to the surface of the soil and trigger germination. Use a spray bottle on the mist setting to just get the surface wet.

Sprouts forming and looking light yellow due to no light exposure.

A filled tray, ready to be exposed to light and green up.

8. Put them under lights. Microgreens respond very quickly to light, turning from yellow to bright green in a day. Read about setting up your microgreens lighting in Using Grow Lights (see page 72).

9. Water when necessary. Give your trays a lift every few days, and if they're feeling light, heavily mist them with water again.

10. Harvest. You'll have microgreens ready to eat in eight days to one month, depending on the variety. Harvest when they're a few inches long using sharp scissors or small grass shears. Grab the top of the microgreens, cut about 1 inch (2.5 cm) above the soil line, and put them in a bowl. Be careful not to get the blades in the soil.

11. Store your microgreens. Pea microgreens store well in a mesh produce bag. Others do well in closed plastic bags. Store microgreens bags in a high-humidity crisper. It's best to use your harvest as fresh as possible—within 3 to 5 days, ideally—for best taste and nutrient density. Wash microgreens just before you use them.

12. Clean up your tray. Most microgreens are finished after one cutting. Your spent microgreen medium and roots are great additions to vermicomposting and hot composting bins (see page 112).

6. Place an empty tray on top. Most seeds don't need light to germinate. Because we're sowing seeds on top of the soil, the darkness and the weight of the tray simulates the conditions of seeds being covered in soil. Putting a little more weight on top forces the seeds to struggle a bit in germination, so they'll grow stronger than if they were just left to germinate with no outside force. Put the prepared seed tray on the bottom tray.

7. Check for germination. Every day or two, look under the top tray for signs of germination. The seeds will push up from the soil and may push up the tray, too. You'll see yellowy-looking stems and know it's time to remove the top tray.

SET A TIMER

EPIC TIP

How often do we, with the best intentions, not set a timer because we're sure we'll remember when it's time to do something? From taking laundry out of the washing machine to checking herbs on the dehydrator, we've all made this mistake. Set a timer on your phone when soaking microgreens seeds prior to planting. I recommend soaking them for several hours to overnight, depending on the variety. If you leave them to soak too long, they can rot, and then you've wasted a lot of valuable seed.

MY FAVORITE MICROGREENS

My favorite microgreen of the moment varies, but this is a standard go-to list of what I like to have available most of the time.

Wheatgrass

Wheatgrass is a popular microgreen for juicing. The seed has a hard seed coat and needs to be soaked for 6 to 12 hours before seeding for good germination. Use 8 ounces (227 g) of seeds per 1020 tray. Wheatgrass microgreens grow very quickly—1 inch (2.5 cm) or more a day—and are ready to harvest fast. These are one of the only microgreens that you can cut and come again, meaning you can harvest from the same plants more than once by letting them regrow.

Arugula

Baby or micro-arugula is super flavorful—a quick and easy way to add a bit of a peppery taste to salads. You want just ¼ ounce (7 g) of seeds to cover a 1020 tray, no soaking required.

Dunn Peas

Pea microgreens are so tasty. Soak these seeds for 4 hours before sowing. Give them space in their soaking bowl, because they'll end up absorbing water and swelling a lot. You need 5 to 8 ounces (142 to 227 g) per 1020 tray.

Black Oil Sunflower Seeds

There are hundreds of types of sunflower seeds. Black oil sunflowers are the best for microgreens. Soak 8 ounces (227 g) of seeds for 4 hours. The colder the water, the better for sunflowers. As the sunflowers grow, their hulls will stay on the leaves until you brush them off. This is an extra step you have to take before harvest, because eating the hulls on the sunflower shoots isn't appetizing.

Salad Mix

Salad mix microgreens are super easy. They come in different varieties, so you can get a spicier mix, with mustards and arugulas, or a milder mix that's heavier on lettuces. Seeding rate depends on the varieties in the mix—let's say it's around ½ ounce (14 g) of seeds per 1020 tray—and these don't need soaking.

SPROUTS

Sprouts and microgreens are often confused for one another. They're actually quite different. Sprouts are grown from seeds that have been soaked in water and germinated without any growing medium. We eat the whole sprout: seedling, seed, and radicle—the first root. (Microgreens seeds are sometimes soaked but not sprouted, and we don't eat the microgreen seed and roots.) Like microgreens, sprouts are fast growing, packed with nutrition, and easy to grow in small spaces. If all you have to work with is 6 square inches (39 cm²) of counter space, you can grow sprouts. Some popular sprouts are mung bean, alfalfa, broccoli, and radish.

Safe Sprouts

Search online for information about sprouts, and you'll be confronted with link after link about food-safety risks. It's important to understand the risks of producing your own food while not being scared away from it.

The food-safety issues stem from seeds contaminated with *Salmonella* or *E. coli*. It's believed that contamination happens during seed production, long before they get into your hands. The pathogens are odorless and invisible, so you can't look at a packet of seeds and know whether they're safe. They become a problem in growing sprouts because abundant nutrients, high levels of moisture, and heat generated in the sprouting process pretty much ensure the bacteria will grow. Then we eat the whole sprout—seed and all.

If you're worried about food safety in sprouts, you can cook them before you eat them. Also, use common sense in not eating sprouts that look or smell like they're going bad.

Sprouts are even easier to grow than microgreens—no soil required!

START SPROUTING

Materials

Clean, sterilized glass jar

2 tablespoons (26 g) alfalfa seed

Water

Permeable lid (either a sprouting lid or cheesecloth and a rubber band)

Sprouting Instructions

1. Fill the jar. Put your seeds into the jar, and add water to half full. Secure the lid.

2. Let it sit. Put the jar in a dark place, like a kitchen cabinet, for 4 to 6 hours.

3. Strain, rinse, repeat. Strain the water through the lid, then rinse and strain again. Let this sit overnight. Repeat the straining and rinsing twice a day until you see the sprouts developing green leaves.

4. Eat up. When the sprouts start developing green leaves, remove them from the jar, rinse them one more time, and enjoy them with your next meal.

Pea and onion sprouts are some of the most popular sprout options.

An abundance of nutrition in just a few days' growth.

EASY HYDROPONIC SYSTEMS

Hydroponics is a means of growing plants without soil. You have a medium that supports the plant roots, plus water and nutrients circulating through the system to feed the plant. This is a little counterintuitive, because when we're growing plants in soil, they do poorly with waterlogged roots. In hydroponics, the water is oxygen-rich, so the roots can breathe, essentially.

Hydroponics uses less water than soil growing, because while there may be lots of water in the system, the water continually recirculates. Hydroponics also encourages fast plant growth, because the plants are under grow lights and in ideal temperatures for hours longer than they'd typically be in sunlight, and the right balance of nutrients are available to them in the water-nutrient solution.

All kinds of plants can be grown hydroponically. In small-space settings, you can do well with greens and herbs. If you have a little more room to grow, tomatoes and cucumbers are fun crops.

Full confession: My first experience with hydroponic growing didn't go well. The cucumbers tasted bad—not just bad, they were disgusting. I learned a lot with that crop, specifically that you have to stay on top of the nutrients in the water.

Hydroponics isn't the easiest way to grow food, but it is efficient, and once you get the hang of it, you can get a lot of food from a small space in no time. I've stuck with this growing method because I like the science of it, I love that I don't have to deal with weeds, and there are fewer pests and diseases to manage.

In this section, I'll walk you through some of what I've learned without getting super detailed. I get into the nitty gritty of hydroponic growing with a whole chapter in my first book, *The Field Guide to Urban Gardening*, including plans for four different types of hydroponic setups.

Deep Water Culture

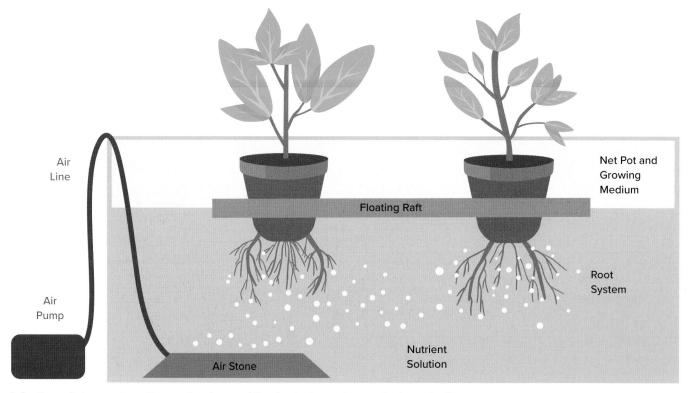

A floating raft deep water culture system is one of the simplest ways to grow hydroponically.

Shredded coconut coir.

Fine gravel.

Seedling starter plugs.

Growing Media

The hydroponic equivalent of soil is called growing medium. This material supports the plant itself, holds the roots, and, depending on the system, absorbs nutrients and water for the plant roots. You have several options for growing media, including these popular choices:

Perlite is a volcanic mineral. It's those white flecks in potting mix that look like little pieces of polystyrene. It's porous and holds onto water and nutrients for plants. It's lightweight enough that it needs to be mixed with another, sturdier growing medium. Look into the sources of perlite, as this is a mined mineral.

Rockwool is probably the most well-known hydroponic growing medium. It's kind of wild to think that rockwool is made of rock that's been melted, spun into fibers, and compressed into cubes. Rockwool has great water retention. You need to use caution when working with it because the fibers and dust can irritate your eyes and lungs.

Coconut coir, also called coco peat, is ground up coconut husks. It holds a lot of water and is easy to work with. You get coconut coir in dry, compressed bricks, and when wet, the medium expands exponentially, like a compressed sponge.

Growstone is sustainably manufactured from glass that's disposed of in landfills. Because it's porous and rough, it's hard to clean between uses and more often needs to be disposed of.

Expanded clay pellets are very popular because they're reusable. These drain and dry out faster than some growing media, and a full, wet tray of expanded clay pellets gets very heavy. Sourcing for clay pellets should be considered, too, as it's a mined material.

Starter plugs have become easier to find as hydroponic growing has become more popular. These are pre-shaped growing medium mixes that you place your seed into and plug into your hydroponic system.

Expanded clay pellets.

Rockwool cubes.

Water and Nutrients

As soil health is paramount in traditional gardening, getting the water-nutrient solution right is essential in a hydroponic garden. The inputs you provide will dictate the balance of nutrients your plants will receive. Mother nature can't help you here.

A top question asked by beginner hydroponic growers is whether tap water will work in your system. The answer is: It depends. Find out from your water utility what chemicals they use to treat the water. If it's just chlorine, you can fill your reservoir and let the water sit for 24 hours so the chlorine dissipates. If you also have chloramine in the water, you need to add a Campden tablet, which is a sulfur-based compound that removes chlorine and chloramine.

Check your baseline pH. The ideal pH is 5.5 to 6.5 for most plants. More acidic or more basic than this and your plants may not be able to uptake nutrients. The nutrients themselves can change the pH of the water, and most commercially available liquid nutrients come with a pH buffer to counteract this.

Check your electrical conductivity (EC) and total dissolved solids (TDS). Water itself doesn't conduct electricity, but as you add nutrients to water, it becomes more conductive. Measuring TDS tells you how many milligrams of dissolved solids you have per liter of water. This is expressed as parts per million (PPM). This chemistry 101 refresher is important because when you measure the EC and TDS in the water, you can determine how concentrated your nutrients are. Having too high a concentration can lead to burned plants; too low, and your plants will languish.

Add your nutrients. Rather than try to formulate your own nutrient ratios—especially when you're just starting out—look for liquid nutrients that are already formulated for the stage of plant growth you're working with. Follow the instructions provided, including the testing protocols for the pH, EC, and TDS discussed above.

You can get a home kit to test the total dissolved solids (TDS) of tap water.

Types of Hydroponic Systems

There are multiple ways to grow hydroponically. This is something that I really appreciate about this high tech natural way of homesteading: It's flexible enough to fit a lot of living situations. Here are the main categories of hydroponic growing:

Wicking systems are the most basic hydroponic system. No air pumps or water pumps are needed here. Nutrients and water are brought to the plant roots using wicking action between the water reservoir and the growing medium. Smaller plants—basil, lettuce, spinach—do great in wicking systems because they don't need a lot of water or nutrients.

The simplest example of a wicking hydroponics setup is a 2-liter (70 oz) bottle garden. You might've even made one of these in science class. Plant a new one every few days, line them up on your kitchen counter, and you can regularly harvest your own fresh greens and herbs.

The Kratky hydroponic method is another that doesn't need pumps or electricity. Kratky systems require a reservoir for the water and nutrients, a net pot and growing medium to hold the plants, and your plants. That's really it. Once you've set it up, this system is practically hands-off. Cherry tomatoes, peppers, leafy greens, cucumbers, and more can be grown here.

In the reservoir, you're setting up your plant with all the nutrients it needs for its lifecycle. Without pumps and channels, this system doesn't recirculate.

Back to the idea that plant roots need oxygen to survive, you have to wonder how plants in the Kratky method can keep their roots submerged throughout their growth cycles. It's science. As a plant takes up water and nutrients, the water level in the reservoir gets lower. The plant's roots will continue to stretch and grow to reach the nutrients, like a drinking straw. Meanwhile, the air gap is expanding between the bottom of the plant and the top of the water. The plant roots get their oxygen from this gap, and all is well.

A Simple Wicking Set-Up in a 2-liter Bottle

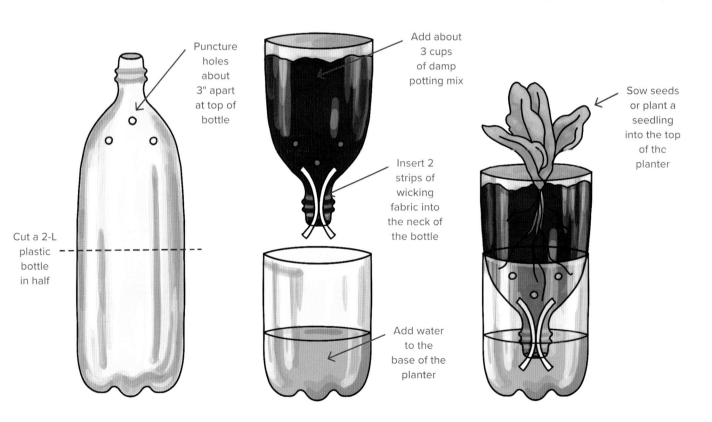

Cut a 2-L plastic bottle in half

Puncture holes about 3" apart at top of bottle

Add about 3 cups of damp potting mix

Insert 2 strips of wicking fabric into the neck of the bottle

Add water to the base of the planter

Sow seeds or plant a seedling into the top of the planter

The Gardyn hydroponics system is an off-the-shelf option with automated watering, lighting, and more.

Deep-water culture (DWC) systems are also easy, straightforward hydroponic systems for small, fast-growing plants. Your plant sits in a net pot filled with a growing medium, and the pot sits in a reservoir of water and nutrients. You use an air pump to oxygenate the water. This is simple enough to DIY and requires little maintenance.

Nutrient film technique (NFT) is a more complex hydroponic system and is more typically found in commercial operations. There's no reason a high-tech homestead can't use NFT, though.

Your setup looks like this: Plants are grown in net pots filled with growing medium, just like in a DWC system. The pots are suspended in a channel with nutrient solution running through it. Unlike deep-water culture hydroponics, only the roots are submerged in the solution. When the solution reaches the end of the channel, it drops back into a main reservoir and recirculates. NFT hydroponics requires little growing medium but more maintenance to keep both an air pump and a water pump running.

Ebb and flow, also called flood and drain, hydroponics acts as the name implies: A tray of plants in growing medium is "flooded" with nutrients throughout the day. Ebb and flow uses a reservoir for water and nutrients, an air pump to keep the solution oxygenated, a water pump to flood the tray, and a timer to run the water pump.

Ebb and flow is a recirculating system, allowing gravity to return the solution to the reservoir after it percolates through the tray. Because you're filling whole trays with growing medium, this uses more growing medium than other systems.

Tower hydroponic systems look very similar to the soil-based grow towers—you'll read about those below—allowing a couple dozen plants to grow in a few square feet of floor space. These vertical systems have a water reservoir at the bottom with a simple water pump and hose to circulate through the central column. With a tall setup like this, having the system on a level floor is important, otherwise one side of the tower will get more water than the other and you risk having water spill over the edge of the reservoir.

You can build your own hydroponic tower using PVC pipes and a 50-gallon (189 L) barrel, or go the high-tech modern route and purchase one that's outfitted with LED lights and smart technology. These self-contained units can adjust light, water, and nutrients to deliver exactly what your plants need, and you can program them using an app on your phone.

Aeroponics requires a more specialized design than other hydroponic systems, and these setups can have faster results. Plant roots are fully exposed to the air and constantly, or nearly constantly, misted with nutrient solution. An aeroponic system requires a growing medium to suspend the plants from, a water line with water misters, a water-nutrient reservoir, and a water pump to keep the system running.

EPIC TIP

BECOME A POLLINATOR

Fruiting plants need some kind of pollination to set fruit. Outside, we have bees, butterflies, and the wind to do that for us. Indoors, it's going to be up to you to see to it that the blooms get the attention they need. Many of the crops you're likely to grow indoors are self-pollinating. Tomatoes, peppers, and strawberries fall into this category. For them, helping with pollination is as simple as shaking the flowers once a day for two or three days to mimic the wind so they release their pollen.

If you are growing something like zucchini, which has male flowers and female flowers, this takes a little more time. Learn to identify a male flower, which has a straight stem and a stamen full of pollen, and a female flower, which has a bump on the stem just below the flower and a stigma. Brush them against one another, or use a small paintbrush to remove pollen from the male flower and transfer it to the female flower.

VERTICAL GROW TOWERS

Vertical grow towers give you dozens of planting spots in just a few square feet of floor space by going up instead of across. This vertical gardening makes so much sense for space-squeezed homesteaders. There are two types of grow towers: hydroponic towers and soil-based towers. You read a little about the hydroponic options previously. In this section, I'm hitting the basics of the soil-based grow towers.

These towers have multiple tiers for growing. Each tier has multiple growing sites, so you can grow a whole garden's worth of diversity in one tower. From the side, each growing site looks like its own container, but the whole tier is connected by soil and water movement.

I recommend having casters or wheels on the bottom of the tower to make it easy for you to rotate the tower so each plant receives the light it needs. Alternatively, you can purchase grow light rings or vertical light strips that integrate with the vertical grow tower model that you purchase. This adds an expense to your setup but takes some of the work out of balancing each plant's light needs.

The GreenStalk Garden is my automatic leafy green machine.

Smart Vertical Planting

Many plants do well in vertical grow towers: strawberries, bush beans, lettuce and other leafy greens, peppers, determinate tomatoes, new potatoes, herbs, and more. Of these, non-fruiting plants are the easiest to grow indoors. You won't be able to produce anything that gets too large, like shrubs.

With all of these plants growing in a tower, you run the risk of taller plants at the bottom blocking the light from the plants growing above them. The growing sites are offset from one another, but there's only so much room to offset this concentration of plants. Think about this when you're putting the plants into the tower, placing taller plants higher on the tower and shorter plants lower.

Start with transplants, rather than seeds. It's better to have more control over your seed starting conditions than what a vertical grow tower will give you. Plant seedlings into the tower after they develop their first true leaves.

Growing Mix

Think of your grow tower as a giant plant pot, because that's what it is. You can use the same soil mix here as you would in any other container. Give it a refresh once a season, and add more as needed. Refer back to Chapter 2 for my advice on container growing mixes (see page 35). You don't have a ton of depth to the growing sites in a grow tower, so fill each tier with soil nearly to the top to give plant roots the maximum amount of room to grow.

Watering the Grow Tower

There's a basin for water at the top of the tower. This means you don't have to water each growing space individually and accidentally spill water all over your floor in the process. Just add water to the basin at the top, and it seeps down through the layers. Because of this seeping action, nutrients can trickle through the layers as well as water.

Top watering forty-two growing sites at once is the pinnacle of watering efficiency in the garden.

USING GROW LIGHTS

Lighting is the biggest challenge your plants will face when growing indoors. It's hard to replace the exact balance of the light spectrum that comes from the sun. Today's grow lights put out an impressive range of light wavelengths at a high density without a lot of energy consumption. This is great news for the plants you're cultivating indoors.

One area where grow lights offer an advantage over natural sunlight is the length of light you can offer to your plants. Sunrise and sunset have no bearing over plants grown under grow lights. You can keep your plants under lights for 12, 16, or 18 hours, maximizing their growth potential.

Let's get into the basics.

Types of Grow Lights

Technology is constantly evolving, and it's possible there will be something new by the time you pick up this book. This list goes from most the basic to the most high-tech indoor grow light options:

T5 fluorescent lights are the shop lights you might have grown up with in your parents' basement. These are typically sold in 2-foot (60 cm) or 4-foot (120 cm) lengths, so you can put together a grow-light setup for your small space pretty easily. Fluorescent lights run cooler than a lot of others. The lightbulbs can be swapped out for those with different color temperatures, which is great for giving plants the spectrum they need based on their stage of growth.

Find the bulbs in normal (NO), high (HO), or very high output (VHO). I recommend the HO bulbs for a good balance of light and heat.

This system integrates a seed-starting tray with a lighting system that elevates as your seedlings grow.

Ceramic metal halide (CMH) is a high-intensity discharge (HID) light. It puts out a well-blended spectrum of light that will give your plants the wavelengths they need throughout their whole growth cycle. The bulbs have a high color-rendering index (CRI), mimicking the color of sunlight, rather than the artificial look of other lights. This is good for the plants but also means they put out UV light that requires proper eye protection when you're working around them.

CMHs are more efficient than a T5, and while they're also more expensive, they're becoming more affordable. These lights last almost as long as LED lights. One last downside to CMHs: They run hot.

Basic LED light units are made up of many light-emitting diodes rather than one single bulb, so you can create a custom light blend of reds, blues, and whites. LED lights use less power per watt and run cooler than fluorescent lights.

In LEDs, you'll get what you pay for. You can find inexpensive LED lights, but you might not get the best performance from them. The lower-end LED lights don't put out as much light as CMHs.

Grow Light Placement

You want your light to have the right footprint to reach all your trays or plants equally. This will save you from having to rotate plants and trays to get their fair share.

The closer you can get your light to the top of the plants or trays, the better. Microgreens and seedlings, especially, are light hungry, and they will expend all their energy growing toward the light and getting spindly. Don't get your lights so close that they burn the plants, though. This is an important thing to watch with the lights that put out a lot of heat, like the CMHs.

Some of the indoor vertical grow towers offer halo-style LED lights or hanging vertical light strips for their setups. These lights attach to the grow tower at intervals that allow each plant to get its share of light exposure.

LED lighting is all most home growers need.

Fancier "quantum board" LEDs are pricey but highly effective.

The Epic Greenhouse, where I start seeds, grow tropical plants, and find peace-of-mind at home.

GREENHOUSES

Greenhouses are pretty sweet, if you have the space and resources for them. They offer an indoor growing space that's as close to being outdoors as possible. In a greenhouse, you can start seeds earlier in the season, overwinter sensitive plants, grow food in-ground year-round, and more.

There are lots of homesteaders who are getting by without greenhouses. In fact, I still don't have one at the Epic Homestead as of this writing. I'm lucky enough to be in San Diego, where it stays relatively warm and relatively sunny most of the year. Building a greenhouse is on my to-do list, but it's not at the top.

A homestead greenhouse can be as basic or as epic as you'd like, ranging from a 5-square-foot (0.5 m²) zip-up plastic model to a state-of-the-art, custom-built glass-walled structure. Choose one that has a supplemental heat source or one that's only heated by the sun. Install automatic vents and fans, overhead sprinklers, and heated tables, or work with it as the basic, useful structure that it is.

In Chapter 1, I touched on where to locate various pieces of your homestead puzzle. It goes without saying that the greenhouse needs to be in full sun. With 8 to 10 hours of direct sun, you're maximizing the sun's rays to warm and grow your plants in this small structure. Orient it lengthwise from east to west to catch as much of the sun's path across the sky as possible.

Tinted glass panes cut UV light and insulate the greenhouse.

You want to have a water source close enough to make watering plants convenient, and if you're running heat, fans, or automated vents, put it close enough to electricity.

Your greenhouse can be a sizable investment. Even if you're DIYing with mostly reclaimed materials, you'll spend a couple hundred dollars. Looking into greenhouse kits, the simplest lightweight greenhouse kits like those that Jacques the Garden Hermit uses will cost less than $1,000 USD. More substantial kits will run a few thousand dollars. Use the best materials you can afford for the inside and outside of the structure so it lasts.

Greenhouse Glazing

The glazing refers to the greenhouse exterior material. Glass gives the greenhouse a classic look but is by no means the only option available today.

Glass allows the best transmission of direct sunlight, but it's not insulative. Heat will be lost through the glass as the air around it cools. Double-paned glass is more insulative but not as much as other options.

Polycarbonate is a durable, clear, ridged plastic. It'll withstand some impact from falling branches, and it's more insulative than glass. The double-wall polycarbonate will keep the inside warmer during colder nights and a five-wall even warmer than that. Polycarbonate plastic can be bent and shaped for more greenhouse styles than straight-edge glass.

Polyethylene sheeting is the thick, clear plastic you see on a lot of market-gardener greenhouses and high tunnels. You can find this sheeting in varying thicknesses. The thicker, the better for durability. One sheet of polyethylene will net great heat during the day but won't hold onto it as the air cools. Covering a greenhouse with two poly sheets allows you to run a fan that inflates the layers like a balloon and does a great job of insulating the structure.

Poly sheeting is less expensive than glass and polycarbonate, but it will need to be replaced every few years. It's prone to tearing or puncture, like when you get too close to it with a weed eater or debris blows into it in a storm. There are lots of plans for DIY Quonset-style poly tunnels online, as well as kits you can buy.

The Insides

The greenhouse is a moist place, by design. Think about this as you're choosing materials to outfit the interior. Coated metal won't rust or mold. As far as wood goes, plywood and pressboard are poor choices. Cedar is probably the most rot-resistant wood choice. Plastic shelving isn't a great idea, because it'll become brittle and break down when it's exposed to this much heat and sunlight.

A thermostat triggers this intake vent to open when it gets too warm.

The exhaust fan triggers at the same time as the intake vent to vent hot air out of the top of the greenhouse.

Cooling and Ventilation

The idea behind a greenhouse is to provide a warm and sunny place for your plants, but small spaces made entirely of windows or plastic will heat up incredibly fast. This is great for plant growth, but too much sun and heat can quickly dry out soil and cook seedlings. The humidity in a greenhouse is ideal for mold and fungal growth, as well, making air circulation important for plant health. You will want to get a handle on this during installation.

In places that get a lot of sun, or if you plan to use your greenhouse in the peak of summer, shade cloth can help create an environment more friendly to plant life and human life. Cover the south side of the greenhouse with 60- to 70-percent shade cloth to keep the structure cooler. This poly-woven fabric is designed to let some sunlight in—the percent number corresponds to how much shade it provides—while giving the greenhouse a break from intense sun rays.

In greenhouses with peaked roofs, you can take advantage of passive cooling with roof vents—no electricity needed. Quonset-style greenhouses benefit from tall windows or doors at both ends of the structure for cross-ventilation. Roll-up sides are nice, too, on days with a breeze.

Fans can help with circulation and ventilation, as well. You can purchase special greenhouse fans, and even a simple house fan can go a long way in moving air. These can be set to come on and turn off as temperatures rise and fall.

Heating

Having a warm space is the whole point behind having a greenhouse. You can get creative in capturing and keeping heat in a greenhouse setup.

The most energy-efficient way to provide supplemental heat to a greenhouse is to fill dark barrels with water. They'll soak up the heat and slowly release it as the air around it cools. If you have space for 55-gallon (250 L) drums, those are ideal, but even 5-gallon (23 L) buckets will do.

Once I switched to greenhouse-started seedlings, their health and germination consistency increased dramatically.

You can run a moisture-safe space heater in the greenhouse, too. Those warnings on space heaters that tell you not to plug them into extension cords are real: This is a fire hazard. A space heater in the greenhouse will only work if you can plug it into an outlet right there.

The big greenhouses and market farmers heat their greenhouses with propane heaters, radiant tables fed by outdoor furnaces, and more. This is probably more than you need as a homesteader, but the options are there, if you want them.

Next, let's step back outside and into the orchard to discuss growing and tending various fruiting trees and shrubs on the homestead.

LET TECH CONTROL THE GREENHOUSE

EPIC TIP

Set a digital thermostat to turn on heaters and fans. Hook up the thermostat to an app on your phone that allows you to set an alarm in case things go wrong, because no technology is perfect. If you put all your trust into the idea that a heater will come on and it somehow doesn't, you want a manual backup plan.

I don't get into greenhouse sprinkler systems here, but they're also great candidates for using tech controls.

This fifteen-tree citrus hedge was planted two years ago and is starting to produce an abundance of rare, delicious fruit.

chapter four

A PRODUCTIVE ORCHARD

ORCHARDS ARE A HOMESTEAD COMMITMENT. You're planting trees now that will be around for decades, and they'll change the landscape as they grow and take shape over time. You're envisioning the future when you're working with an orchard. These trees start out as spindly sticks in the ground, then grow and fill out. In just a couple of years, an area that looks sparsely planted will become a thriving green space full of fruit.

Backyard orcharding—even more than gardening—needs thoughtful planning. Like most of our favorite food-producing plants, fruit trees want to be in full sun. Unlike these other plants, the trees create shady areas in the yard. These shady areas could be wanted, if you're creating microclimates, and could also be unwanted, blocking sun from other plants. Below ground, tree roots can uproot concrete and interrupt pipes and utility lines. Orchard planning leaves a lot to consider.

As part of this homestead commitment, orchards also require patience. A citrus tree might produce a few fruits in its first year, but the real yield comes with time. Apples and pears can take as long as ten years to fruit, depending on the variety. And to keep all of them producing well, you're in store for ongoing care.

Scan for video

WHAT TO GROW

Deciding what to grow in your homestead orchard is as much about which varieties as it is about which fruits. I love citrus—like, really love citrus—so about half of my orchard is in citrus trees. I have two varieties of apples that are adapted to growing in my climate. I also have a peach, a nectarine, some pomegranates, and a couple of avocados. This list will probably grow, and it goes to show that even in a small backyard orchard space, you can plant what you love.

Pick Varieties for Your Climate

Apples are notoriously hard to grow in Southern California, but both of my trees are producing well. I selected the varieties—Dorsett Golden and Anna—because they can grow in warm climates.

Citrus, on the other hand, is right at home here but wouldn't make it through the winter outdoors anywhere colder than Zone 8. Citrus trees can also be grown most anywhere in containers that can be brought inside for the winter.

Chill hours. Look for the number of chill hours the fruit variety requires. Chill hours tell you how many hours the plant needs to accumulate in the 32°F to 45°F (0°C to 7°C) range in order to flower and set fruit. Here in San Diego, I plant "low-chill" fruit varieties because we don't have many hours that fall within that temperature range. A homesteader in a northern growing zone would want to plant varieties that require high chill hours.

Inspecting a Bearss lime tree, one of my most productive trees at the homestead.

Microclimates. In Chapter 1, I talked about taking advantage of your microclimates. This is a great strategy for orcharding, where just 5°F (3°C) can make a difference in fruit production. An easy microclimate trick is to give yourself another zone or half zone by planting a tree a few feet off a south-facing wall. The wall will absorb heat and reflect it back to the tree, keeping it warmer through winter and bringing it out of dormancy sooner in spring.

Native plants. Native trees are naturally well-suited and adapted to your area—no microclimates or overwintering indoors needed. Look to your region's native fruits for relatively easy-to-grow orchard options.

Think About Space

The amount of space you have is your most limiting factor in what you can grow. Within the space you're working with, it's time to prioritize your favorite fruits and make hard decisions. Keep reading for my thoughts on dwarf and semi-dwarf trees, tree spacing, and pruning. You can probably fit more trees in an outdoor space than you thought possible.

A Frederick passionfruit planted in front of my rainwater cistern, which will grow in and provide beauty and fruit, as well as cool down the cistern.

Look for Self-Fertile Varieties

Some trees require mates for cross-pollination. They're considered self-infertile. This is the case with avocado, persimmon, apricot, apple, sweet cherry, plum, and others. Within those, some even need to have a mate of the same cultivar for fruit production. Whenever you grow self-infertile trees, know that you're giving up twice the valuable orchard space to that variety.

My Eva's Pride peach, a prolific producer.

Thinning these peaches out will give me less, but better, fruit.

Self-fertile trees, on the other hand, don't need a mate to make fruit, so you can get by with just one tree of these varieties. Partially self-fertile trees can make fruit on their own but are more productive when they have a mate.

Research varieties before overcommitting your space.

Or Plant Grafted Combinations

There is a way around the need for self-fertile varieties. You can purchase combination grafted trees: Multiple varieties or even multiple fruits grafted onto and growing from the same rootstock. Get your Frost peach, Puget Gold apricot, Hardired nectarine, and Stella cherry all on the same tree; your Meyer lemon, Minneola tangelo, and Valencia orange right there together; your Yoinashi, Hamese, and Mishirasu Asian pears just a branch apart—you get the idea.

It might be harder to find the exact combination you like, but these combination-grafted trees will save you from having to plant multiples of the same fruit in your small space.

Choose Your Root Stock

Assuming you're using grafted trees, getting the fruit you want isn't just about the variety but the rootstock, too.

Grafted trees are common because fruit trees are choosy in how they reproduce. Fruit trees don't typically breed true, so if you plant McIntosh apple seeds, you might not get what you think of as a McIntosh apple tree. Cuttings from fruit trees are hard to root, so that's not a great way to propagate them, either. Grafting, instead, takes a cutting—a scion—from the fruit-producing part of the intended variety and joins it with the root system—the rootstock—of another, usually hardier, variety. By the time the tree ends up on your doorstep, it's grown together well enough that you can hardly tell there are two plants there.

Rootstock choices include those that will produce dwarf vs. standard trees, varying levels of winter hardiness, whether it can handle poor soil drainage, and resistance to pests and disease. Look for a rootstock that matches the conditions on your homestead.

Decide Between Bare Root and Potted Trees

As you're tree shopping, you'll find some trees that come to you "bare root"—just swaddled in a protective wrap—and some that are potted. Each has their place, and you may find one works for you now and the other is more appropriate for your next tree purchase. Learn the difference in the Bare Root Trees vs. Potted Trees chart on page XXX.

Manually grafting different fig varieties onto a single tree.

BARE ROOT TREES VS. POTTED TREES

	BARE ROOT	POTTED
Lightweight; easy to handle	X	
Limited planting window, during dormancy	X	
Generally less expensive	X	
Available all year long		X
Can be up-potted if you're not ready to commit to a planting spot	X	X
Can get rootbound and suffer transplant shock		X
Gets started developing a healthy root structure immediately upon transplant	X	
Wider range of cultivars available		X

A grafted fruit tree with the grafting scar visible on the right side.

The Epic Homestead has over twenty varieties of fruit planted close together, a feature of backyard orchard culture.

HOW TO GROW FRUIT TREES

It's one thing to have acres and acres to devote to fruit production. It's something else altogether to try to grow a variety of fruits on a small homestead. Go ahead and ignore that voice that's saying you don't have enough space. Backyards, front yards, and containers are all places you can grow fruit trees.

Backyard Orchard Culture

The phrase "backyard orchard culture" was coined by Dave Wilson, of Dave Wilson Nursery, in Hickman, California. I've learned a lot about fruit trees from Tom Spellman at this nursery, and the backyard orchard culture concept captures exactly what I'm working on at the Epic Homestead. Backyard orchard culture focuses on getting a good amount of a variety of fruits from a small space. By maintaining smaller plants and planting them in a compact way, you're giving up huge harvests, but you're getting more variety—just what you'd need for a homestead. Here are the basics of making backyard orchard culture work.

Choose smaller varieties. This first piece of advice for backyard orchard culture is maybe not the most accurate. Yes, you want smaller trees, and dwarf and semi-dwarf cultivars are meant for small-scale orcharding. It's also true that you can have a great harvest from full-sized trees that you keep pruned to a small size. There's more on this in a minute.

Plan for varieties with successive ripening. Successive ripening allows you to space out your harvest over weeks or months. There are two ways to plant for successive ripening: Plant a few different varieties of the same fruit that have varying ripening windows, or plant different fruits altogether that will ripen throughout the length of the season.

Apples are a great example of planting for successive ripening. If you've ever been to an apple orchard, you know you can expect apples throughout the growing season, but different varieties are available at different times. Let's say you have a zone 6 homestead orchard, and you want to eat fresh apples from June through November. Your apple tree planting might look something like:

- Lodi, ready late June
- Dayton, ready early August
- Gala, ready late August
- Golden Delicious, ready mid-September
- Fuji, ready mid-October
- Granny Smith, ready late October

As opposed to focusing on just one fruit (apples), if you wanted to have a variety of tree fruits over the season, you could have a successive ripening planting that follows the When to Expect Tree Fruit chart on page 87.

Plant with tighter spacing. You're breaking all the rules of traditional orcharding here. Tight tree spacing reduces the trees' growth, keeping them compact. It also allows you to pack in more varieties of fruit than you could using traditional orchard planting methods.

A great example of using tight tree spacing is with the avocado trees that my garden assistant, Jacques Lyakov, and I planted. We put them just 2 feet (61 cm) apart in my backyard, and I intend to treat them as if they were one tree as they mature. Large-scale avocado plantings call for 15 to 25 feet (4.6 to 7.6 m) between trees, so when I say this is breaking all the rules of traditional orcharding, I mean it. Dave Wilson Nursery recommends planting as many as four fruit trees in a 4-foot-by-4-foot (1.2 x 1.2 m) raised bed and putting in hedgerows with 3 feet (91 cm) between trees.

Prune aggressively. You get to choose the height and shape of your trees with your pruning. This is intimidating at first, but you'll get the hang of it. Aggressive pruning also improves airflow around the fruit trees, which will help curb disease in this tight spacing.

Two pomegranate varieties grown "as one" for more consistent harvests.

The first pomegranate fruit forming at my homestead.

The gorgeous, waxy, red flower of the pomegranate tree.

Loosening the root ball of a Hass avocado to prepare it for planting.

Haas and Fuerte avocadoes planted in the "two in one hole" method.

This chart looks at typical ripening dates for tree fruit in USDA Zone 6. This is an estimation of dates, which will depend on the weather, the variety, and your particular microclimate.

FRUIT	RIPENING DATES
Cherries	Early to mid-June
Apples	Late June to late October
Peaches	Late June to mid-August
Apricots	Early July to late August
Nectarines	Early July to late August
Serviceberries	Mid-July to late August
Plums	Early August to late September
Asian Pears	Early August to late October
Paw Paws	Mid-August to early October
Pears	Mid-August to mid-October

Eighteen months of growth on the citrus hedge, lining the north side of the property.

2-D Planting

Pruning into espalier and other 2-D shapes is a good option to add a visual interest to your homestead and to maintain fruit trees in truly small spaces. A blooming apple growing along a wall or a pear loaded with fruit trailing on a trellis is something to see. Trees that produce on long, lateral branches are the best for this planting strategy: apples, pears, figs, peaches, cherries, and pomegranates.

Citrus Hedge

Citrus trees grow more like shrubs than they do trees. This is a nice feature for growing citrus on urban and suburban homesteads: You can plant a citrus privacy hedge.

The citrus hedge at the Epic Homestead creates a sort of oasis vibe, with much of the north side of my property lined with these evergreen trees. Planted 3 to 4 feet (91 to 121 cm) apart, the greenery from one tree will grow into the next, providing a green wall that also throws off hundreds of delicious—and a few rare—fruits. In a few years, these trees will completely block my view of the neighbors and the street on that side.

The backyard homestead is a tapestry of different food-producing planting layers.

Food Forest-Style Planting

I mentioned food forests in Chapter 3 as a strategy to grow vegetables in a way that mimics nature's own growing style. Don't let the "forest" word scare you away: A food forest can be as small as a raised-bed garden or as big as an actual forest. You can even set up a food forest with your container plants. This planting style brings in everything from ground cover and low-growing plants to taller annuals, perennials, shrubs, and trees.

The permaculture community identifies seven layers to a food forest:

- **Canopy layer.** Tallest layer, provides shade and cover for lower layers. If you have deciduous trees and shrubs in your canopy, they'll provide natural mulch ground cover with their falling leaves.
- **Understory layer.** Slightly smaller plants, probably still shrubs or tall-growing vegetative plants, like sunchokes.
- **Shrub layer.** Berry bushes and similar-sized plants.
- **Herbaceous layer.** Vegetable plants and herbs, whether annual or perennial.
- **Ground cover layer.** Strawberries, low-growing herbs, other trailing plants.
- **Root layer.** The roots and rhizomes that are working to pull nutrients into plants, the soil life creating healthy underground ecosystem, and the like.
- **Vertical layer.** Vining plants, like passion fruit, beans, and butterfly pea vine.

Food forests need to be planned so each plant is exposed to the amount of light it needs. The north side of the planting will get the least amount of light, shaded by the plants to its south. The tallest plants belong along the north side, and your shortest plants along the south side.

A lush container-only edible front-yard garden still counts as a homestead!

Container Growing

Strange as it seems, fruit trees are doable in containers. It's a great strategy for growing fruits in colder climates. Container growing a small orchard also works when you don't have the right spot to plant your tree in ground yet or when you don't have a place to plant it in ground at all.

The best candidates for container growing are columnar trees, dwarf trees, and slow-growing trees. Fast-growing trees, like figs and mulberries, are not a good idea for containers—unless they're dwarf varieties. The full-sized fast growers will outgrow your biggest pot in the first season.

Choosing containers. Larger containers are better because they allow the roots to stretch out. You want a 25-gallon (95 L) pot at minimum. You'll need to repot the tree into a slightly larger container every other year, so don't start out potting your tree in the largest container you can manage.

Consider the weight of the container, if you're growing on a balcony or another weight-bearing structure. Grow bags are nice container options for trees, because they're as lightweight as containers come, and they have handles for moving these big plants.

Whatever container you choose, go for one that has good drainage. Trees in containers need regular watering, but they don't want to sit in saturated soil.

Container care. Top off the container planting with mulch so the container stays cooler in summer and warmer in winter. Mulch will also help keep moisture from evaporating from the soil.

You're in control of all the nutrients that the plant receives in a container. It can't bury its roots deep into the earth to extract what it needs. Plan to fertilize with a balanced fruit-tree fertilizer regularly.

There's a whole section on growing fruit trees in containers in my book *Grow Bag Gardening*.

Jacques Lyakov, Epic Gardening's resident Garden Hermit, and I planted an orchard in his backyard to provide his chickens with shade and protection from aerial predators. With figs, a mulberry, a loquat, and a lemon, he has a good mix of evergreen and deciduous trees to provide a variety of fruit as well as year-round flock protection.

This is a symbiotic relationship, as chicken droppings are beneficial to orchards. Letting them run here is a more food-safe plan than turning them loose in gardens because tree fruit isn't touching the ground as it grows. For as small as they are, chickens can be destructive. Their natural scratching and pecking behaviors are great for aerating soils, but too much of this can damage root structures. To provide some protection from the chickens and to raise the trees out of the heavy clay soil, Jacques planted each tree into a Birdies Tree Surround raised bed frame. He made a sort of collar out of hardware cloth and covered that in mulch to further keep the chickens from scratching and pecking directly on the root area.

Check out the whole process of planting the chicken orchard in a video on the Epic Gardening YouTube channel.

Scan for video

EPIC TIP

DOLLIES FOR YOUR TREES

If your living situation allows, put your container trees on dollies. They make it easy to wheel the tree in and out of the house, greenhouse, or garage as the seasons change without having to lift heavy pots.

Using a tree surround, Jacques created an edible orchard that also protected his hens from predators.

Contrary to popular suggestions, you don't need to dig a huge planting hole for fruit trees.

Materials

Potted tree

Shovel

Soil amendments, as needed

Mulch

Water

Paint-on protective coating (I like IV Organic 3-in-1 Plant Guard)

I planted the first tree at the Epic Homestead with the help of my friend Cameron Akrami, of The Busy Gardener, and I still use the advice he gave me. This project is less about needing specific materials and more about the tree-planting technique that works for me. This project covers planting a potted tree. See Planting a Bare Root Tree, page 88, to read about the difference.

Tree Planting Instructions

I. **Prepare the area.** Move away any mulch in the planting area. When it comes time to refill the planting hole, you don't want to fill it with mulch—just soil—so get the mulch out of the way now.

2. **Dig the hole.** Dig a hole slightly larger than the circumference of the potted tree. Don't dig too deep. Depending on your soil, you'll plant with 2 to 4 inches (5 to 10 cm) of the potted tree above the soil line. The heavier your soil, the more shallow you want to plant. This helps water drain from around the root zone.

Roughing up the soil to break the roots out of their spiral growing pattern helps a tree settle into its new home.

3. Prepare the hole. If you have heavy clay soil like mine, rough up the interior edges of your planting hole. As your tree roots grow and hit the edge of the hole, you don't want them to think they've hit a brick wall. Don't be tempted to add amendments to the planting hole. When you first put in a tree, it needs time to establish roots and become stable before it can start growing. The tree will find what nutrients it needs from the native soil. You can top dress with fertilizer later.

4. Prepare the tree. If you're planting a potted tree, you have to first get it out of the pot. You have the option of slicing a flimsy plastic pot down the side with a knife and peeling away the pot. If you want to keep the pot for another use or the pot is too sturdy for slicing, give it some taps on the sides to loosen the soil away from the edge. Grasp the tree at the base of the trunk, turn over the pot, and the tree and soil should slide right out. Resist the urge to pull the tree straight out of the pot by its trunk—you can damage the root structure that way.

Assuming your tree is not rootbound, scratch up the surface of the soil mass to break its form.

If the tree is rootbound, you'll see a mass of roots spiraling around the bottom. Pull apart these roots. It's not a big deal if you break them—they will regrow.

5

I planted this citrus higher than the native soil level to help with drainage as it establishes itself.

6

Backfilling the hole with native soil—no amendments added when planting out.

7

Covering with mulch and watering in for an hour or so.

7. Mulch. Cover the planting area with 3 to 4 inches (8 to 10 cm) of mulch—hardwood is ideal.

Give the tree 4 to 6 inches (10 to 15 cm) of mulch-free space around the trunk to breathe. If you mulch tight against the bark, the mulch could hold moisture against the trunk and cause rotting and diseases.

8. Water. This initial watering helps to settle the soil and give the tree roots good soil contact. The tree's root systems are much larger than those of vegetables. Run a trickle of water for a half hour or an hour—the deeper the watering, the better.

9. Paint on a protective coating. I like IV Organic 3-in-1 Plant Guard to protect the trunk from heat, potential diseases, and pests. This protection is especially important for late-spring- and early summer-planted trees. Avocados and citrus seem to be the most prone to sunburn. Coat all of the exposed areas of the trunk and main branches.

10. Water consistently. While trees don't want to sit in soggy soil, they do need regular watering as they're getting established. Water two or three times a week for the first couple of weeks to get the root system going, then once a week after that.

You might even intentionally cut apart a few. You can also flare the root ball: Build a mound of dirt at the bottom of the planting hole, and spread the bottoms of the roots apart so they're reaching in different directions around the mound. Breaking up this ball of roots allows them to grow deeper into the soil rather than continue to circle around themselves. There's one important exception here: Avocado trees do not want to have their roots bothered. Just plant them as they are.

5. Settle the tree. Place the root and soil mass into the hole. The soil mass from the tree's container should be a few inches (cm) above the top of the hole.

If you're planting a grafted tree, orient the graft scar toward the north. You don't want the sun blasting that unprotected point, especially on citrus trees. Be sure the graft point remains above the soil line.

Stand above the tree so you're looking down from the top, and be sure the trunk is growing straight up. The tree will continue to grow in the direction it's planted.

6. Fill the hole. Backfill around the tree with the soil you removed from the hole. Tamp down the soil as you're filling. When the hole is full, mound up the native soil and slope it away from the tree to give the area better drainage. Give the whole planting area a good pat, or step on the freshly filled hole, to pack in the soil.

PLANTING A BARE ROOT TREE

Bare root trees have a few different planting needs than potted trees. Read about how I plant my potted trees in Plant a Tree (page 92) and compare.

- Bare root trees can only be planted while they're dormant. You have to get them in the ground when you receive them.

- Soak the tree in a bucket of water for a few hours before planting.

- Dig your hole a good deal wider than the roots are to give the roots room to reach as they start growing. They'll start establishing their root structure right away.

- Leave a cone-shaped mound of soil in the center of the hole so the roots have something to sit on.

- Plant a bare root tree deeper than you would a potted tree, at the same depth it stood in the ground at the nursery. The trunk flare—where the trunk meets the roots—should be level to the ground surface.

- Backfill the hole carefully, being sure to tamp out any pockets of air. Then water in the tree as described in the previous project.

Bare root trees are cheaper but only available in early spring. They also take longer to grow out.

EPIC TIP

GET A MOISTURE METER

Overwatering is the biggest mistake new orchardists make. Your homestead's trees will thank you for not watering too much with the help of an inexpensive gadget called a moisture meter.

You can find moisture sensors that will connect to irrigation systems, but all you really need is a portable, handheld device. More accurate than guessing how wet the earth is, you insert the probe to its depth and get a reading of the moisture and pH at that soil level.

PRUNING

Orchards are pretty hands-off after getting trees established, except for the work of pruning—and there's little work more nerve wracking than pruning. You're taking off whole parts of a living tree—sometimes cutting off immature fruits—with the hope of getting more fruit and making the tree healthier in the long run. It's almost counterintuitive.

I prune for aesthetic value as well as tree health. The Epic Homestead is not only about producing as much as possible in this small space: I want a yard that looks nice, too. Rather than just let the trees run as they will, pruning helps the tree maintain a balanced structure.

Pruning citrus trees is different from pruning peach trees, which is different from pruning apple trees. Let's look at my pruning strategies.

The Three Ds

Pruning starts with observation. Your orchard trees are something you might pass by without paying too much attention. Set aside some time to really "see" your trees and evaluate them through this simple framework.

For all trees, you're looking for the three Ds: dead, diseased, or damaged. Cut off any dead branches, any disease, and any damage. With the three Ds, it doesn't matter the time of year. The point is to prevent issues from spreading to other parts of the tree and to other trees.

Remove diseased and insect-infested plant material from your property. Don't put this in your compost. It might make great bonfire material, though!

Dead/diseased stems

Damaged stems

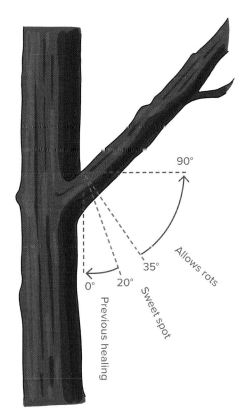

90°

Allows rots

35°

Sweet spot

0° 20°

Previous healing

This is how to make a proper pruning cut angle.

Below the Graft

Grafted trees, remember, are trees with two or more different trees spliced together. The bottom portion, below the graft point, is called the rootstock. Any shoots growing from the rootstock will produce fruit from the rootstock variety, not from the variety that you carefully chose and planted. These rootstock shoots need to go. Cut them off at the trunk as soon as you notice them.

SUMMER PRUNING

There are different opinions on when to prune your orchard trees. I use summer pruning to keep trees at a manageable size. Shorter trees are easier to harvest from, less likely to shade out plants around them, and fit into the backyard orchard culture system. Summer also tends to be a drier time of year, which is better for pruning. Other orchardists will tell you to prune in fall or winter, when the tree has stopped growing for the year, which is the strategy to follow if you want large trees. But pruning while the tree is still in a growth stage encourages more compact growth rather than sprawling growth.

Pruning top growth in summer is scary but helps keep your plants at a manageable height.

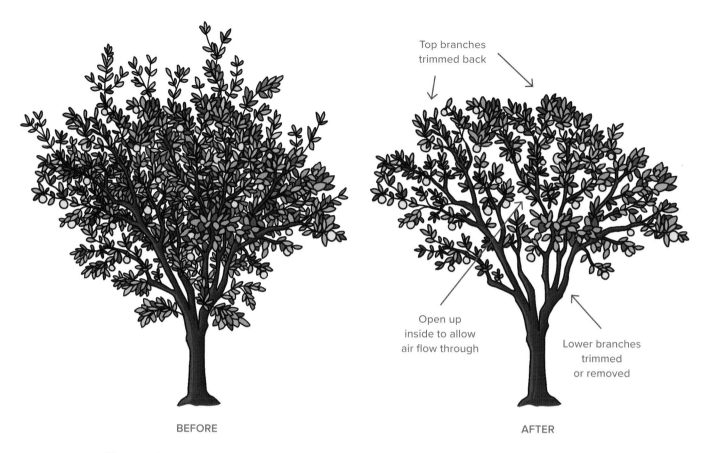

Top branches
trimmed back

Open up
inside to allow
air flow through

Lower branches
trimmed
or removed

BEFORE

AFTER

Citrus can be pruned at any time and is more about shape and height management than anything else.

Pruning Citrus Trees

Citrus trees are not deciduous. They don't lose their leaves in the winter.

They like to have a nice, tight canopy—more like a bush or a hedge than a tree. Come late summer, citrus trees send up tall stems that need to be cut. Find the height you want, and cut the stem just above a node at a 45-degree angle. This angle makes sure water cannot sit on the open cut, reducing the potential for rotting or disease.

Citrus is also prone to "water sprouts," which are vigorous shoots coming out of latent buds. You'll be able to tell a water sprout easily, as it grows quickly and vertically from lower points on the tree's trunk. Remove these completely.

In the first year after planting citrus trees, they'll sometimes try to fruit. Here's the hardest part of pruning: You have to cut off that fruit. Leaving it will take energy from the plant that should be going into getting a root system and healthy structure established. Sacrifice the fruit in your first year to gain fruit in the years to come. If you can't bear to remove all the fruit, I've had success leaving one to two, maximum, on the tree so I can still enjoy a taste of what's to come.

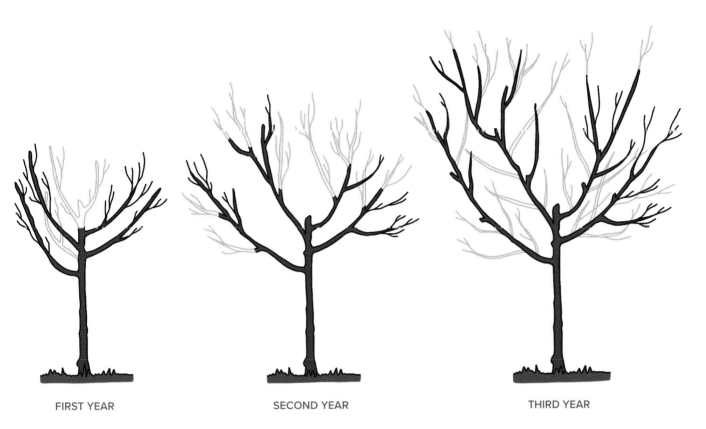

FIRST YEAR SECOND YEAR THIRD YEAR

Stone fruit trees should be pruned to have open canopies with branches that can support next season's fruit.

Pruning Stone Fruit Trees

Stone fruit trees—peaches, plums, cherries, and others with the large, stone-like pits—are deciduous, so they lose their leaves in the winter. They need an open canopy, unlike citrus's shrubby growth habits.

Stone fruit trees with long branches can't support their fruit as well as tightly pruned trees. The first year my peach tree produced, I had more fruit than expected. If I hadn't pruned the tree well the winter before, that fruit would have weighed down and broken the branches. Even with that pruning, I had to remove about fifty immature fruits because the branches were so heavy!

To prune the vertical-growing stems, locate the main stem. Cut the stem to the height you want to keep the tree—generally 6 to 8 feet (1.8 to 2.4 m) in a backyard orchard. Use a 45-degree-angle cut. For fast-growing peaches and nectarines, you may find yourself cutting back new growth by as much as half.

To prune lateral shoots—the shoots coming off the vertical stems—count a few buds out from the vertical stem, and cut at a 45-degree angle. The bud right below will grow and continue the branch the next season, and the ones below that will fruit.

Look for branches that are crossing one another, and cut off these branches at the main stem. While trees look hardy, fruit trees are prone to fungal issues, especially in humid climates. Trimming these keeps the canopy open to allow for good airflow that will reduce fungal growth.

Low-lying off-shoots of already low-lying branches are also fine to cut off. Those shoots don't add to the fruiting capacity of the tree, so you don't want the tree to keep putting energy into them.

FIRST YEAR   SECOND YEAR   THIRD YEAR

Pome trees need a main trunk to support their growth over the years.

Pruning Pome Trees

Apples, pears, and quinces are in the pome fruit family. These trees are also deciduous and lose their leaves in the fall. The healthiest pome trees have a vase shape: one trunk and multiple main branches growing off at vertical angles. This allows the interior of the tree to receive light and airflow.

Remove any stems that are growing off the main branches toward the middle of the "vase" to keep that structure open. Also remove the branches that are angled downward or horizontal. These will not be strong enough to hold the weight of fruit. Branches growing upward at a 60-degree angle are ideal.

Cut back the tallest branches on pome trees to the height you want in your orchard.

TYPES OF PRUNERS

Go ahead and look up orchard pruners, and you'll find options ranging
from $10 USD to several hundred dollars. The amount you want to spend is up to you.
As long as you have at least one pair of pruners and one pruning saw that you like,
you should be all set for pruning in your homestead orchard.

This is a quick guide to the types of pruning tools out there:

Bypass Pruner

If you could only choose one type of pruner, get a good pair of bypass pruners. A sharp pair can make clean cuts on branches 1 inch (2.5 cm) or less in diameter. The mechanism of bypass pruners slices through the plant material and does not crush it.

Long-Handled Bypass Lopper

These tools use the same cutting mechanism as the bypass pruners. Their long handles give you more leverage for cutting larger-diameter branches and allow you to reach branches in awkward spots. You can find these with telescoping handles, too, to give you more reach and leverage.

Bypass Pole Pruner

Maybe you're not as tall as I am, at 6-foot (1.8 m) plus, or maybe you're tending an orchard from a wheelchair. There are tools that make fruit-tree care accessible for everyone. These bypass pole pruners are handgrip-activated. Position your pruners and squeeze the hand grip to activate the cutting mechanism at the end of the pole handle. You can get into small spots and tall spots with a pole pruner.

Ratchet Pruner

For branches too wide for bypass pruners or too tough for your grip, ratchet pruners have a ratchet action that helps you slice through the wood one increment at a time. They're so good for harder pieces. I'm surprised by some of the things I've been able to cut with these.

Power Pruners

A rechargeable battery powers these pruners to do the work for you. Instead of squeezing grips, you just pull a trigger, and your branch is cut.

Pruning Saw

Folding pruning saws, or quick saws, have blades serrated with sharp teeth. I love these—they make such short work of tree branches. They fold up for easy storage and to keep those blades away from your fingers when you're not using them.

Battery-Powered Pruning Saw

The battery-powered pruning saw essentially looks like what it is: a mini-chainsaw. I say that quick saws make short work of tree branches, but these battery-powered saws take care of even larger branches on older trees. They have the safety mechanisms and maintenance requirements of regular chainsaws.

10 COMMON ORCHARD PESTS AND DISEASES

ISSUE	LOOKS LIKE	FOUND ON
Apple Scab (*Venturia inaequalis*), fungus	• Black, scabby lesions on leaves and fruit	Apple, crabapple, and pear trees
Fire Blight (*Erwinia amylovora)*, bacteria	• Drooping, shriveled flowers • Young growth wilts, turns gray-green, and bends downward in a hook shape characteristic of fire blight. • Dark, shriveled fruit • Sunken, dark cankers on bark	Apple, pear, and serviceberry trees; raspberry brambles
Brown Rot (*Monilinia fructicola*), fungus	• Wilted, brown, dead flowers and shoots • Soft, rotten areas covered with fuzzy fungal spores on fruits	All stone fruits; plums are less susceptible
Huanglongbing (also known as HLB or Citrus Greening), *Candidatus Liberibacter asiaticus*, bacteria spread by Asian citrus psyllid insects	• Asian citrus psyllids on plants • Damaged or reduced tree growth. • Twisted, curled leaves and characteristic leaf notching • Sooty mold growth on sap excreted by psyllid nymphs • Asymmetrical yellowing of leaves on one limb or in one sector • Smaller, bitter fruit • Partially green, lopsided fruit • Premature fruit drop • Chronic cases: Sparsely foliated trees with small leaves that point upward; extensive twig and limb dieback; eventual death • Symptoms may not appear for 2 or more years after infection.	All citrus
Spotted Lanternfly (*Lycorma delicatula*), insect	• Groups of spotted lanternfly adults, egg clusters, and nymphs on trees • Oozing wounds on bark that lead to wilting and death of branches	Apples, nectarines, plums, peaches, cherries, grapes, blueberries, and more

CAUSED BY	CONTROL WITH
Weather	• Plant resistant varieties and rootstocks. • Rake and destroy fallen leaves. • Prune for air circulation. • Spray fungicide.
Warm, humid, rainy spring and summer weather; often after hail, winds, or rain that damages trees. Bacteria seeps out through cankers as sweet, sticky liquid that's spread by insects, moved by splashing water, and transferred by hands and tools.	• Plant resistant varieties. • Prune before the blossom period. • Avoid heavy pruning. • Prune from 8 to 12 inches (20 to 30 cm) above diseased tissue, and discard diseased limbs. • Prune to maintain air circulation. • Use antibiotic or copper sprays.
Rainy weather during bloom	• Prune infected parts. • Spray fungicide just before or after blooming; apply just before ripening if weather is wet.
Asian citrus psyllid nymphs feed on new citrus tree shoots and leaves. They remove sap and leave behind a salivary toxin and the HLB bacteria as they feed.	• Invite parasitic insects—including lady beetles, syrphid flies, lacewings, parasitoid wasps, and minute pirate bugs—to control Asian citrus psyllids. • Remove ant populations around citrus trees, as they protect the psyllids because they eat their honeydew (waste). • Apply soil, systemic and foliar insecticides. • Destroy infected trees. • Respect quarantine zones, if you live in a state that has them, like California. • Report suspected HLB infection or Asian citrus psyllid infestation to your state department of agriculture.
The insects suck sap from stems and leaves, which weakens and damages the plants. They excrete honeydew waste, which causes molding and brings in other insects, such as wasps, hornets, and ants.	• Respect the quarantine zone, if your state has one. • Report sightings to your state department of agriculture. • Spray or inject systemic insecticides. • Maintain plant health always.

ISSUE	LOOKS LIKE	FOUND ON
Plum curculio (*Conotrachelus nenuphar*), insect	• Tiny, ¼-inch (6 mm) plum curculio beetles on plants; cat facing, ⅛-inch (3 mm) crescent-shaped cuts, and small larvae exit holes on fruit • Premature fruit drop • On peaches, there's often a gummy material around the wound.	Apples, and stone fruit
Codling Moth (*Cydia pomonella*), insect	• ½- to ¾-inch (1.3 to 1.9 cm) long mottled gray moths that blend in with tree bark • White to light pink "worms" (larvae) with a dark brown head inside the fruit • Holes in the fruit filled with reddish-brown, crumbly frass • Larvae overwintering in thick, silken cocoons under loose bark scales or debris around the tree	Apples and pears
Scale, insect—many species	• Scaly, armored pests ranging in color from light orange to brownish-purple to snow white • Leaf yellowing and drop • Decreased fruit production • Green spots on fruit • Limb dieback; occasionally death of the tree • Black, sooty mold on the honeydew waste	All citrus, blueberries, figs, mulberries, persimmons, plums, quince, and others
Aphids, insect—many species	• Swarms of tiny gray, green, black, yellow, or brown pear-shaped insects • Masses of aphids gather inside curled leaves • Leaves turn yellow, twist, and die • Leaves with dark spots, plaque, and mold	Most orchard plants
Stink bugs, insect—several species	• Shield-shaped, armored-looking insects in various shades of brown, gray, and green • Dimples or cat facing on fruit • White, pithy areas that turn brown when fruit is peeled, concentrated near the stem end • Small, brown, tear-shaped insect droppings	Most orchard plants

CAUSED BY	CONTROL WITH
Adult plum curculios do their damage when nighttime lows begin to reach 60°F (10°C). The female adult cuts a hole in the fruit and deposits an egg in a small cavity. She cuts a crescent-shaped slit beneath to prevent the egg from being crushed by the developing fruit. Larvae hatch and burrow into the fruit about 5 days later. When fully developed, larvae leave the fruit through holes with no frass or webbing.	• Apply insecticides at petal-fall or first-cover stage for apples and the shuck-split and first-cover stages in peaches and cherries. • Pick up and destroy damaged fruits as they fall.
Female codling moths lay 30 to 70 eggs on fruit, nuts, leaves, or spurs. After hatching, larvae bore into fruit and tunnel to the core. They can infest 20 to 90 percent of fruit in an orchard.	• Plant varieties less susceptible to codling moth damage, such as early maturing fruits. • Remove and destroy fruits with frass-filled holes. • Clean up dropped fruit. • Cover fruit with cotton tie-string bags or No. 2 paper bags 4 to 6 weeks after bloom to exclude moths. • Use insecticide sprays.
Scale insects suck sap from the plants and excrete sticky honeydew waste, which coats infested plants. Dark fungi called sooty molds grow in the honeydew.	• Attract parasitic insects, like parasitic wasps and ladybird beetles. • Trim off and dispose of infested foliage. • Spray horticultural oil before new growth begins in late winter or early spring.
Aphids suck sap from leaves, damaging young shoots. Their sticky honeydew waste attracts ants. Many species are the biggest problem in late spring, when temperatures are not yet hot.	• Observe plants to catch infestations early. • Bring in insects to eat the aphids: such as lady beetles, lacewing larvae, and soldier beetles. • Prune and dispose of curled leaves. • Keep the inner canopy pruned to make the habitat less suitable. • Use a strong hose to spray the aphids off the plants. • Spray a water-soap solution, insecticidal soaps, or insecticidal oils.
Stink bugs feed directly on fruits, rather than on plant parts. The trees are undamaged, but the fruit is severely damaged.	• Don't grow stink bugs' favorite plants around orchard trees: mullein, mustard, and dock, plantain, milkweed, mallow, morning glory, thistles, vetch, and broadleaf plants. • Spray with insecticides. • Spray leaves of plants with kaolin clay.

No homestead is complete without some kind of composting system—a way to break the organic waste you generate down into fertilizer for your future harvests.

chapter five

COMPOSTING

YOU COULD SAY there's something rotten about homesteading: It's compost.

You hear this term, compost, all the time when talking about food scraps, garden cleanup, soil amendments, and chicken waste. Compost is as simple as it is valuable. You're left with compost when organic matter is decomposed by organisms and microorganisms. This same action occurs in the soil, as you incorporate organic materials and leave plant roots in place after a crop is finished. The material breakdown happens faster in a compost pile because you can create the ideal conditions for decomposition. Here, nearly before your eyes, what was once whole pieces of plants, handfuls of straw, and kitchen vegetable scraps turns into fine, loose, dark, soil-like compost.

Composting on a homestead of any size just makes sense. You're growing plants, you're cooking and preserving food, and maybe you're keeping chickens—all these things produce nutrient-rich organic waste that you can turn around and put back into your garden. Compost feeds the soil microbes with available nutrients, improves water retention in garden soil, and improves heavy soil.

Composting at home not only saves you from having to pay to dispose of homestead waste, but it saves you from having to pay to bring in soil nutrients. The best part about composting is it'll happen whether you set it up properly or not. There are ways you can make compost happen faster and more efficiently—some with little effort on your part— and we'll explore those in this chapter.

Scan for video

I built a five-bay hot composting system to keep up with the volume of organic matter the Epic Homestead produces.

If brown materials are scarce, bales of hay make a cheap compost additive.

FOUR KEY COMPOST INGREDIENTS

Great compost follows a recipe. You can't have compost without these four ingredients:

- **Carbon.** Carbon-rich materials are also called "browns." These include uncoated cardboard, crumpled up uncoated newspaper, dried leaves, loose straw, wood mulch, and dried up plants. The browns provide energy to the microorganisms working in the pile.
- **Nitrogen.** Nitrogen-rich materials are the "greens." These are fresh, disease-free plant material, food scraps, untreated grass clippings, and whatnot. The greens are like protein for the microorganisms.
- **Air.** Hot composting is an aerobic process. The organisms need oxygen to thrive. Bokashi composting is the only anaerobic compost method you'll read about on page 127.
- **Water.** The organisms also need water to survive. Compost piles need more water than you might expect, because as organisms are breaking down the materials, they're creating heat, and that heat is drying out the pile. There is such a thing as too much moisture, too, which you'll read about soon.

A pile of miscellaneous garden scraps used as green materials.

Pill bugs are wonderful decomposers—in the right amounts!

Compost is a product of a colony of organisms and microorganisms. Too small to watch without a microscope, microorganisms are the powerhouses of the compost pile. Bacteria do most of the decomposition and heat generation. Fungi get to work on the tougher plant pieces, and once those are broken down, the bacteria can step in to finish them off. Protozoa are microscopic organisms living in the water droplets in the compost. These ingest bacteria and fungi, which is partly why you don't want your compost to be too wet.

You may find insects in the compost pile, too. Sowbugs, spiders, earthworms, millipedes, and centipedes move in when bacterial and fungal populations are lower. They break down organic matter, though not as efficiently as the microorganisms. If you see a lot of these bugs, it could mean your compost is nearly finished, or it could mean your pile could use some help to call the microorganisms back to work.

Carbon-to-Nitrogen Ratio

Compost-ingredient ratios are less about hard math and more about paying attention. The ideal ratio of browns to greens is 30 to 1, meaning if you put in 1 pound (0.5 kg) of rotten tomatoes (a nitrogen source), you need to add 30 pounds (14 kg) of brown leaves (your carbon source) to balance the pile. Greens tend to weigh more and be more available than browns, making a 30:1 ratio hard to maintain.

Jacques Lyakov, my garden assistant, has had success with a 50:50 ratio, which is more manageable from both a mindset and a resources standpoint. With a 50-carbon-to-50-nitrogen ratio, when you add 1 pound (0.5 kg) of a green material, you need 1 pound (0.5 kg) of a brown material.

I've seen hot compost work with a 2:1 brown-to-green ratio: for example, 1 pound (0.5 kg) coffee grounds (nitrogen) with 2 pounds (0.9 kg) wood mulch (carbon).

This C:N explanation is simplified. Digging deeper, you realize each item you put into the compost bin has some carbon and some nitrogen. Nothing is higher in nitrogen than carbon, and the amount of carbon or nitrogen in each item will vary depending on how fresh or dried it is. For example: Brown leaves are 60:1 C:N, grass clippings are 20:1.

As you can tell, this ratio can become really complicated. The closer you get to a true 30:1 ratio, the more efficiently the microbes will work to break down your compost pile. It's up to you how nerdy you want to be with your compost management.

Your compost will tell you when your browns and greens are out of whack. Too much green material makes the pile smell bad and be too wet. Too many browns will cause the pile to not heat up or break down, and it'll be very dry. The fix for an unbalanced pile is easy: Just add the opposite ingredient to remedy your problem.

Watch the Temp

The temperature inside the compost pile can help you gauge its oxygen and moisture level. A long-probed compost thermometer is an easy way to monitor the temperature.

Within a day of building your compost pile, the temperature should read 120°F (49°C). The temperature you're shooting for is 140°F to 160°F (60°C to 71°C). Different microorganisms will move in based on their ideal temperature. If the compost gets hotter than 160°F (71°C), most will die off or move out.

Compost thermometers help you track the stages of your compost.

With a little composting success, you'll see every pile of garden debris as pure gold.

Besides keeping a healthy space for organisms to do their work, a compost pile at the proper temperature will also kill the seeds and pathogens in the pile. Weed seeds, vegetable seeds, and flower seeds remain viable up to temperatures of 160°F (71°C). Managing Manure, see page 129, touches on controlling pathogens.

It tends to be that compost piles run too cold rather than too hot. There are three ways to increase the temperature of your compost pile:

- **Turn the pile to aerate it.** With a compost tumbler, this is simple enough: Just crank the handle. In an actual pile, you'll need a shovel, pitchfork, or compost aerator to manually flip and stir it. Ideally, you'd turn the pile every day. As often as you can—at minimum once each week—is my advice.
- **Have the right amount of material in the pile.** The compost pile should be a minimum of 3 cubic feet (0.08 m³) to get the biological activity needed to produce heat.
- **Add more greens.** Nitrogen-rich materials will heat up a pile fast. You don't want to tip the balance too far and make the pile too hot, so do this carefully.

Watering Your Compost

Water each time you turn the pile to keep the mixture damp enough for the microbes to be happy. If you squeeze a handful of compost, you want it to feel like a damp sponge. If you live in a wet climate, you might have to build a roof over your pile so you can better control the water going into your compost.

HOT COMPOSTING

There are two ways to look at composting: hot composting and cold composting.

Cold composting is the lazy man's compost—throwing all your organic materials in a pile and waiting for things to break down. You'll get compost eventually, and if this is all the time and energy you have for composting, this composting method is better than no composting method at all. I won't spend much time talking about cold composting, because I prefer other methods, and there's really not much to it.

Hot composting, on the other hand, breaks down the organic materials faster and allows you to process more volume of waste. It also requires pile management, with the regular pile turning, carbon-to-nitrogen balancing, and watering covered previously.

EPIC TIP

USE PIECES BIG AND SMALL

It makes sense that smaller particles will break down faster in the compost pile. You can add whole, fibrous plant pieces, like sunflower stalks, and eventually those will break down. When you chop those pieces before tossing them, there's more surface area for the microbes to work on, and they'll break down faster.

A mix between the two is important, as the large pieces will help to create air pockets and allow water to filter through. In time, they'll break down into small pieces and become microbe fodder.

A five-tine compost fork is the perfect pile-turning tool.

Straw or a reliable brown material as a topper helps protect the pile.

Layer Your Pile

To really get hot composting right, you might need a holding bin to keep your materials before they make it to the compost. Layering browns and greens is the best approach to creating prime conditions.

Get your compost pile started as if it were a sandwich:
- The first layer is the soil, where your organisms and microorganisms will migrate from. Think of it as a plate to hold your sandwich.
- Then have a layer of browns. This can be a layer of larger pieces, creating some breathing room at the bottom.
- Greens come next.
- Top this off with another layer of carbon-rich browns.
- Continue this green-and-brown layering for a rich, diverse compost sandwich.

Water each layer well before adding the next. In the next few days, you'll mix up your layers, so don't let perfection get in the way of getting a pile started.

Black gold is starting to form in this pile.

A cross-section of a pile working well.

COMPOST CONTAINERS

While you can always compost in a free-standing pile, there are plenty of contained options that offer multiple benefits. Here are some of my favorites.

Compost Tumblers

A compost tumbler is a plastic barrel suspended above the ground on a stand. This is a nice feature for small-space gardens, because you can still use the space underneath for growing plants, or you could put it on a patio.

The compost tumbler design is genius. The enclosed bin keeps out rodents and other animals. Air holes allow for air circulation, and the tumbling mechanism does your compost turning for you. The downside to the air holes is that as you water the compost and as it rains, the bin will leak water.

The tumbling action keeps the smaller, denser pieces together in the middle to create the heat that radiates out to the rest of the pile.

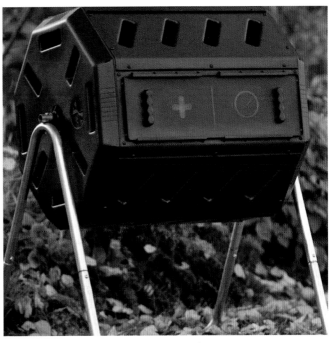

A compost tumbler was my first experience with the composting process—and it was fun!

Marrying an indoor food scrap bin with an outdoor tumbler is the perfect small-space waste solution.

TRY A COMPOST ACTIVATOR

EPIC TIP

Sometimes you follow all the "rules" and your compost just won't get started. Compost activators are like sourdough starters, full of good microbes just waiting to break down some organic matter. You can purchase a compost activator to mix into your pile, but sometimes all you need is a scoop of finished compost. The bacteria and fungi present there can colonize and get to work on the pile you're building.

BUILD A MULTI-BAY COMPOST BIN

If you're not comfortable with power tools, big-box stores will cut lumber for you for a small fee.

Materials

Side panels

8 3-foot (91 cm) 2 x 6 boards

8 3-foot (91 cm) 2 x 4 boards

20 3-foot (91 cm) 1 x 6 boards

Back

6 9-foot (2.7 m) 1 x 6 boards

1 9-foot (2.7 m) 2 x 4 board

Front

8 3-foot (91 cm) 2 x 2 boards

18 32-inch (81 cm) 1 x 6 boards

1 box #8 x 2¾ inch (7 cm) cabinet screws

8 U-posts

A multi-bay compost bin is one of the most effective ways to recycle your garden and kitchen scraps. The three-bay system here is maybe the most common. Add scraps to the same bay every day until it's full. Give the second bay time to break down into compost, and use the finished compost from the third bay in your garden.

At the Epic Homestead, I built a five-bay compost bin system because of the size of my garden. Even five bays aren't enough to hold and break down all the materials when I'm busy turning over garden beds. Let these instructions guide you to build the size of compost bin you need. If you want a five-bay system, just do the measurements to add on additional bays. Likewise, if you have room for just one bay, cut this down to size.

Construction Instructions

1. Measure and cut. If you brought in full length lumber, go ahead and cut it now.

2. Build the side panels. For every panel, you'll make a sandwich with 1 x 6s as the "filling" and 2 x 4s and 2 x 6s as the "bread."

2a. Lay 1 2 x 4 board on the ground parallel to 1 2 x 6 board, about 2.5 feet (75 cm) apart. Evenly space 5 1 x 6 planks running across these boards, with the sides, top, and bottom of the 1 x 6 planks coming flush to the sides, top, and bottom of the 2 x 4 and 2 x 6 that are resting on the ground. The space between the planks will allow airflow through the compost pile.

2b. Top this sandwich with 1 2 x 4 board on top of the planks on the 2 x 4 side and 1 2 x 6 board on top of the planks on the 2 x 6 side.

2c. With everything lined up, put two screws through each top board, through each plank, and into the bottom boards. Start with the 2 x 4 side (the back side) and then move on to the 2 x 6 side (the front).

Measure out the spacing between cross-boards.

Sandwich the boards in place.

Screw the sandwich together with cabinet screws.

2d

All bay sides built. Almost done!

2d. Repeat this with the remaining side panels. A three-bin system will have four total side panels.

3. Build the back.

3a. Stand up the side panels on their 2 x 6 sides so the 2 x 4 side is at the top. Use cinder blocks or wood planks to prop them up, if needed. Space them 3 feet (91 cm) apart.

3b. Attach 1 2 x 4 board across the bottom of all side panels to tie them together. Be sure it's lined up with the edges of the outside side panels.

3b

Lay down the back supports.

3c. Attach the first 1 x 6 board flush to the first 2 x 4, edges still lined up with the ends of the outside side panels. Attach the remaining 5 1 x 6 boards flush to one another and the outside side panels.

4. Flip your bin so it's properly oriented. The part you just assembled should be the back of the bin.

5. Assemble the front panels. These are modular so you can easily take out the finished compost.

5a. Using scrap wood, create a 1 inch (2.5 cm) spacer. This doesn't have to be long—6 inches (15 cm) or so will do.

5b. On the left inside of the first bin, attach 1 2 x 2 board flush with the front of the side panel. Clamps are a big help here.

Back assembled and leveled.

Flip the compost system upright.

Measure the removable front slats.

5c

Install the channel to hold the front slats.

5d

Test fit the slats.

5e

Finish off the front of the system.

5c. Using that spacer, attach 1 2 x 2 board 1 inch (2.5 cm) behind the first 2 x 2.

5d. Repeat on the right side and left side of each bay to create the channel so the front slats can slide in and out.

5e. Drop in each 1 x 6 panel to create the front of the bin.

6. Attach one U-post to each side of each corner of the compost bin to shore up the structure.

6a. With the flat side of the U-post flush to the front left side of the compost bin, pound the post into the ground.

6b. Use two screws to secure the U-post to the bin. Be sure to leave the front-panel channels open.

6c. Repeat with each remaining corner of the bin.

6c

Watering in the first pile of the new system.

Scan
for video

MAKE A PALLET BIN

EPIC TIP

If you have access to pallets, use those to build your multi-bay compost bin. Be sure they're not treated with chemicals that can leach into your compost. Check the letter code stamped on the pallet. Stick with pallets stamped HT, KD, and DB—meaning they have been heat treated—or pallets with no stamp on them, indicating they have not been treated. Do not use pallets stamped MB, meaning it was treated with the pesticide methyl bromide, and EUR, meaning it's an old European pallet that may have been fumigated.

Trash Can Composting

Compost is anything but trash, but composting in a trash can is a pretty good idea. Heavy-duty plastic and metal trash cans with tight-fitting lids are useful all over the homestead, and here especially. Drill aeration holes in the sides of a trash can, and you have a versatile compost bin. Consider these three uses:

- Weigh down the bottom with bricks or a concrete block, and build a standard compost pile inside. Water, add organic materials, and turn this as you would any compost pile.
- Instead of weighing down the trash can, make it a compost tumbler. Secure the lid with bungee cords and roll it around the yard to aerate the compost every couple of days.
- Bury the trash can in the soil as a trench composter, or use it as a buried worm bin. (Keep reading for details about both.)

An aerated trash can is one of the simpler budget compost bin options available.

VERMICOMPOSTING

I can geek out over compost of all kinds, but composting with worms is my favorite compost method. It's interesting to put worms to work to break down your food and garden scraps.

As worms devour their food, they leave behind nutrient-rich waste called worm castings. These worm castings fall to the bottom of the bin for you to collect and use as a well-balanced soil amendment. Worm castings also add microbial life to the garden soil, including, occasionally, a worm—or their eggs!

Success with vermicomposting starts with making your worms comfortable. They need a place to live: some kind of container with air flow. Inside the container, you layer bedding and food scraps, then add worms. Worms are cold sensitive, preferring nothing colder than 55°F (13°C), which is something to consider for northern gardeners.

A downside to vermicomposting is that it's hard to do on a scale that would compost all the waste produced by a big garden. You'd need multiple worm bins to do this, anyway. So vermicomposting might be your sole compost system, if you have a small homestead, or it could complement an additional method.

Worms are your best friend, both in soil and in your compost.

Worm Bedding

Bedding needs to be an organic material, as the worms will consume this along with the food you provide. Coconut coir, shredded uncoated paper, shredded straw, paper towels, and shredded uncoated cardboard are great bedding options. The more diversified sources of bedding, the better to offer different particle sizes and allow airflow. The bedding should be moist but not wet. When you squeeze a handful of it, you want a drop or two of water to fall, but not more than that—like a sponge you just squeezed the water out of. Let the bedding sit for a few days before adding worms so it can start to create an underground, microbe-rich environment that worms like.

Every two or three feedings, add more bedding to keep the nitrogen-to-carbon balance in check and to soak up the moisture from the food scraps. The ideal here is 4 parts bedding to 1 part food.

My favorite mix is 50 percent coconut coir and 50 percent shredded paper.

Overfeeding your worms early on is the fastest way to kill your worm bin.

Worm Food

You're a little more limited in your worm bin inputs than you are in other compost systems. Feed your worms fruits and vegetables, but not citrus, onions, or garlic. They can't break down meat or dairy fast enough to keep the whole system from going bad, so keep those out. Worms also will avoid oils, so skip anything that's been cooked with oil. It's easiest for worms to digest smaller bits of food, so for example, chop up that wedge of cabbage instead of putting the whole thing in there.

Steve Churchill, who owns Urban Worm Company, told me that the biggest mistake vermicomposters make is overfeeding their worms. Aim to feed your worms about 25 percent of their weight each day. And when I say 25 percent, I should also say that this number depends on what you're feeding your worms. Offer less of a food with high moisture content—pumpkin, cucumbers, etc.—or a little more of a food with lower moisture content—banana peels and broccoli stalks.

When you give the worms more food than they can process, the food scraps go rank, and you have a host of problems. As food scraps rot, they create heat, and too much heat can kill off the worms. You can effectively hot compost your worms, if you're not careful—ask me how I know! Now you have even fewer worms to eat what was already too much food. Fruit and vegetable scraps are mostly water, so they can cause things to get too soggy, unless you're adding absorbent bedding to balance this. This moisture also creates conditions that attract flies and everything starts to stink.

Meet the Worms

Red wiggler is the worm species you want to employ. They thrive in rotting vegetation. You might be able to find these in an existing compost bin, but you're better off buying some so you know what you're getting. I started my system with 2 pounds (0.9 kg) of red wiggler worms, and I didn't have to add more after that because the system self-regulated. They'll reproduce when the conditions are right and will hold or reduce their population when conditions aren't ideal.

Worm Bins

Worms aren't all that picky about where they live. Dark, damp, and breathable are their top requirements. Here are a few tested and true types of worm bins:

The Red Wiggler is the go-to composting worm of choice.

Worm tote. Worm tote designs are plentiful. You can purchase a whole kit or make one yourself. These totes can be small enough to sit under your kitchen sink or large enough to need their own space in the garage.

In the realm of homestead building projects, building a worm tote is on the easy side. You can find a video with instructions for how I built a worm bin for less than $30 USD using three plastic totes on the Epic Gardening YouTube

channel. The idea behind the multi-bin worm tote is that you'll end up with a layer consisting of bedding and food scraps in one tote and a layer of worm castings in another. It's a smart system.

Using worm totes, it's just a few weeks before you can harvest worm castings and leachate—what I like to call worm juice, which is the nutrient-rich liquid that will seep out of the system.

Stackable tote systems are easy to build and work quite well for a DIY option.

The Urban Worm Bag is my go-to system for easy worm composting.

You can make in-ground bins from a 5-gallon (19 L) bucket with holes drilled in it to keep costs down.

Worm bag. Also called a continuous flow-through worm bin, these reinforced canvas bags can hold a lot of materials to produce great worm castings. The Urban Worm Bag is my recommendation for a continuous flow-through system.

Unlike a bin system that requires stacking and unstacking, the whole organic-breakdown process takes place in one container. As the worms process the food scraps, natural layers form with the worm castings and leachate falling to the bottom—where you can scoop them out without disturbing the rest of the system. You add more bedding and food scraps in the top, as needed, and you'll get usable worm castings in four to six months.

In-ground worm bin. In-ground vermicomposting is an ingenious way to inoculate your garden beds with soil life and collect worm castings that you can use right there or spread to other beds. An in-ground bin has holes all around the bottom and sides so worms can come and go from the bin to your garden bed. Add bedding and food scraps to this bin just as you would any other worm bin. Having it right there in the garden makes it a convenient place to toss your garden clippings, too.

Like the above-ground systems, you can build your own in-ground bin or purchase a ready-made system.

BOKASHI COMPOSTING

If you can only dedicate a 5-gallon (23 L) bucket's worth of space to composting, you have enough space for bokashi composting. Bokashi—a Japanese compost method—is the heavy-hitter. In other compost methods, we're told not to add dairy products, meat, fats, oils, and bones to the pile. Bokashi can handle all those things.

This is an anaerobic compost method. The other composting methods I write about here all require oxygen for the microbes to thrive and the materials to break down. Bokashi composting takes place in a sealed 5-gallon (23 L) bucket. To this bucket, you layer a grain inoculated with *Lactobacillus* anaerobic bacteria with food scraps. This mix ferments in the bucket. After a few weeks, you won't recognize the contents of the bucket when you open the lid. The next step is to bury the contents in your garden or add it to an aerobic compost bin to finish breaking down.

To get started with bokashi composting, you need to purchase inoculated grains. Other than that, all you need are a brick or something similar as a weight and spacer, a couple of old towels or T-shirts, wax paper, and two 5-gallon (23 L) buckets. One of these buckets needs a lid that seals tight. You can find DIY bokashi bucket instructions on the Epic Gardening YouTube channel.

A negative feature of bokashi composting is the limited space in this 5-gallon (23 L) bucket. If you're composting from your garden as well as your kitchen, you probably need to use an additional compost system to handle all your waste. You can't add big pieces of scraps and waste to the bucket. A potato the size of my fist, for example, would need to be cut up. Also, because bokashi composting relies on a specific strain of bacterium for fermentation, you shouldn't add anything that's moldy because the fungus and bacteria would compete.

Food scraps sprinkled with layers of inoculated bokashi bran.

Scan for video

TRENCH COMPOSTING

Trench composting is just what it sounds like: digging a trench and burying your compost in the soil. It works if your soil life is relatively healthy. Burying food scraps is an age-old means of making compost and feeding the soil. In Chapter 2, you read about some of the garden styles that incorporate trench composting into their design, including hügelkultur and keyhole gardening. This isn't the most efficient means of composting, but it is low-tech and easy. Your scraps are directly feeding the soil as they break down.

If you want to compost meat, dairy, bones, and other things that are tasty to creatures living outdoors, burying them like this isn't the way to go. Pests will dig in your garden to go after the food they want, making a giant mess along the way.

Try a trench-composting experiment: Dig an 18-inch (46 cm) deep trench down the center of one of your beds. Bury your garden and food scraps, and tamp down the soil. Plant a row of crops on both sides. When you water your crops, be sure to water your trenched row, too. Watch how this bed compares to the others in your garden. The scraps will break down throughout the season, slowly releasing nutrients to the plants and soil around them.

A whole trench isn't even necessary here. If you only have space for a hole in the center of a square or round bed, do that instead. You can even sink a vessel, like a trash can, with holes in it into the ground and put your food scraps in there.

You'll be shocked how fast bokashi food scraps break down in trenches.

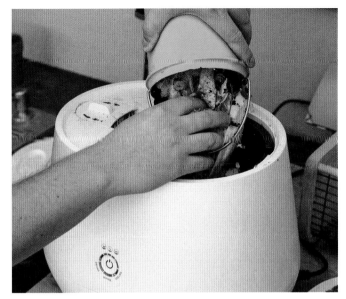
The Lomi is my countertop composter of choice.

COMPOSTING IN APARTMENTS

Having lived in an apartment when I first started exploring the idea of homesteading, I'm here for all the apartment dwellers. Maybe you've lucked into a building with a shared garden and compost bin out back. Odds are better that you haven't. Now more than ever, you have options.

Countertop Composter

Among the many appliances useful to people who, like homesteaders, spend a lot of time in the kitchen, electric composters or countertop composters are pretty cool. They look not unlike an ice cream maker, but instead of churning out delicious, creamy desserts for your family, they turn your food scraps into rich compost for your soil.

This is as easy as composting can get. When you put the scraps from your most recent meal into the bucket, the appliance dehydrates and grinds it. What you end up with is pre-compost, ranging in consistency between veggie chips and soil. The moisture has been removed and the volume significantly reduced, but it's not fully composted. You can add this directly to your garden area or to another compost system to continue breaking down.

MANAGING MANURE

The deposits animals leave behind are full of microbes and nutrients that soil life wants. They're also full of pathogens that you don't want around your food. Working with manure is less of a soil health issue than it is a food-safety issue. Whenever you're composting animal waste—whether that's horse manure from a farm down the road or chicken poop from your own homestead—different rules apply.

To start, you never want to add pet waste or pig manure to your compost. These carry pathogens that your home compost may not destroy.

I'll give a nod to U.S. Department of Agriculture National Organic Program compost guidance here. To safely break down manure, your home compost pile must:

- Start with a carbon-to-nitrogen ratio of between 25:1 and 40:1.
- Be kept at 131°F to 170°F (55°C to 77°C) for at least 15 days and turned at least 5 times during that period.

If this is too much pressure, you can follow guidelines for raw manure, which is to:

- Apply it only to areas growing ornamental plants.
- Add it to the soil at least 120 days before harvesting food that touches the soil, like lettuce, strawberries, and root vegetables.
- Add it to the soil at least 90 days before harvesting food that doesn't contact the soil, like tree fruits, broccoli, and pole beans.

While expensive, countertop systems may be your only option in small spaces.

The countertop composters can handle dairy, as well as chicken and fish bones. Some electric systems will break down bioplastics, like plastic bags made from corn. Larger bones, fruit pits, oils, and sugary foods won't process here.

A major downside to the electric composters is the price. At a few hundred dollars, this system requires an investment that others don't. Also, this is obviously a very small machine, not something that can handle a garden's worth of spent plants. On the plus side, you can break down food scraps in a few hours to a few days, rather than putting them in the trash.

Compost Pickup

Smart and scrappy compost businesses are popping up in cities everywhere. You can sign up to have someone pick up a bucket of your food scraps and leave behind a clean bucket. Often, your subscription includes access to finished compost throughout the year. Look into local compost subscription services where you live.

COMPOST THIS, NOT THAT

	HOT COMPOST	VERMI-COMPOST	BOKASHI COMPOST	TRENCH COMPOST	COUNTERTOP COMPOST
Most vegetable and fruit scraps	X	X	X	X	X
Garlic and onion	X		X	X	X
Citrus	X		X	X	X
Dairy			X		X
Meat and bones			X		Some meats and bones
Eggshells	X	X	X		X
Cooking oil and oily foods					
Moldy food	X				
Cooked food		X	X		X
Clippings and debris from healthy plants	X	X	X	X	X
Diseased plant material					
Compostable plastics (bioplastics)					Some compostable plastics
Uncoated paper and paper towels	X	X		X	
Animal manure	X	X	X	X	
Inorganic materials					

Solar power unlocks a new world of opportunity for building your dream homestead.

chapter six

ENERGY SYSTEMS

WHEN LOOKING AT HOMESTEADING from a modern perspective, energy systems is a place where the "high-tech, natural" concept stands out. Solar energy and energy conservation are a huge focus at the Epic Homestead, because without at least attempting to create energy, I don't think it would be fair to call this a sustainable project.

Around the world, electricity grids are straining under the pressure of increased demand. Rolling brownouts and blackouts are becoming commonplace in some areas. In the United States, we see them in my home state of California and even in Texas. I think it's possible to use a homestead to contribute to energy solutions rather than energy problems. You can use the energy you harness to improve your ability to produce your own food and your quality of life.

With an 8.5-kilowatt system, my solar panels produce more energy than I use. (Incidentally, the median size residential system in 2021 was 7.0 kilowatts, according to Berkeley Lab's Electricity Markets and Policy department.) This overage allows me to live more comfortably, which is partly the point of modern homesteading. I have the freedom to use a timer for the chicken coop door, run the pond pump without worry, automate garden irrigation, and even enjoy my mini-split air conditioning as often as I'd like. Even with slightly less energy generation during the shorter days of the winter months, my solar panels still generate more energy than I consume.

Smart homestead energy systems are not just about energy production. They're also about energy conservation and the benefits that come from big-picture energy thinking. In addition to solar power, which is the backbone of this chapter, I'll cover mini-split air conditioning and heating units, lighting, batteries, and more.

Scan for video

Before I installed my panels, I upgraded my roof from a poorly-done torch-down to a TPO roof.

SOLAR POWER

San Diego ranks number two in solar power per capita among cities in the United States. Not installing solar panels at the Epic Homestead would have been a big mistake.

I started out with a fourteen-panel system that I later upgraded. I now have twenty-five photovoltaic panels on the roof at the Epic Homestead. The cost of the panels was about $20,000 USD. With my house being more than 100 years old, I needed to replace the roof, too, for another $12,000 USD. I would guess most roofs don't have to be replaced—or maybe an area of the roof would just need to be reinforced—and skipping roof replacement would save a lot of money. I was able to get a significant tax break with solar tax credits, making this a large investment but one that made sense.

Making Sense of Solar

There's so much to consider when deciding whether an investment in solar energy makes sense for your homestead. Here are some things I took into account when designing my system.

Financial return on investment (ROI). The initial financial outlay is high, but you'll make up for it over time. Most solar-installation payback periods are six to ten years. Mine should be five-and-a-half to six-and-a-half years. Your ROI period will depend on a number of factors and may be longer or shorter. See Determine Your Solar ROI on page 136.

A suitable climate. What's your area's typical forecast? Solar power harnesses the energy of the sun, so if you don't regularly see the sun, you won't generate as much energy. I've studied the energy-production reports on my energy-system provider's app, and I'm estimating that energy production decreases by 20 to 25 percent on a cloudy day.

Having moderately cloudy weather shouldn't knock out solar as an option altogether. You might just need a larger array or need to supplement your solar output with energy from the grid.

Ever since these twenty-four panels went live on my roof, I've paid $0 for electricity.

Property layout. Another reason solar made sense for me was my relatively flat roof and lack of tree cover. I got lucky when my new neighbor cut down the trees along our property line that were blocking my sun! It was easy enough to orient the solar panels primarily to the west and south for ideal sun exposure. If you have a steeply pitched roof, a house surrounded by tall trees, or a roof with a pitch that doesn't align with sun exposure, you may be able to place the solar panels elsewhere in your yard.

Zoning. I touched on planning and zoning and homeowners association ordinances in Chapter 1. They definitely apply here. Solar panel sizing and placement can be dictated by your city, county, or housing development. Know these rules and the permits required before you shop for a system.

Intangible benefits. Also on the list of considerations could be benefits that are harder to quantify. Lots of people look to solar power wanting to live off grid. There's also the issue of climate change and wanting to live more lightly on the earth, and not drawing energy from nonrenewable sources. Plug in your own values here.

My solar output would have been much worse if these trees were still on the property line.

Unbound Solar, a renewable energy company in the Pacific Northwest of the United States, determines solar-installation payback using this simple formula:

Total system cost minus *Rebates and incentives* divided by *The cost per kilowatt of electricity* divided by *Your annual kilowatt usage* = **The number of years it will take to recoup your expenses**.

Hypothetically, let's say you live in Durham, North Carolina, and your system costs $20,000 USD.

You have a 30 percent federal tax credit (until 2033), taking the cost down to $14,000.

Divide this cost by $0.12 per kilowatt of electricity for 116,667.

Divide that number by an annual kilowatt usage of 16,000 kWh, and you end up with a 7.29-year ROI period.

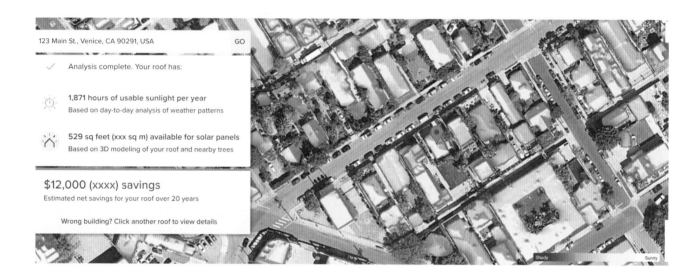

123 Main St., Venice, CA 90291, USA — GO

Analysis complete. Your roof has:

1,871 hours of usable sunlight per year
Based on day-to-day analysis of weather patterns

529 sq feet (xxx sq m) available for solar panels
Based on 3D modeling of your roof and nearby trees

$12,000 (xxxx) savings
Estimated net savings for your roof over 20 years

Wrong building? Click another roof to view details

Net Metering

When I started exploring solar energy, I naively thought that solar energy was going directly from the photovoltaic panel into my house. That's not true. The energy either goes into a battery, which I'll explain in a bit, or it goes back into the grid. Here I am, producing more energy than my homestead needs, and I actually draw my energy from the grid, not from my solar panels. This seems backward, but in the sense of my electricity bill, it all works out.

When it's time to calculate my electricity bill each month, the energy utility looks at the amount of electricity I use and the amount I produce and bills me—or credits me—for the difference. This is called net energy metering.

Net metering allows customers who generate their own electricity to sell it back into the grid. This is the norm in most states here in the United States. In states where net metering isn't mandated, some utilities still offer this perk. The credits I accrue can be used toward an electricity bill I might have in the future.

It makes sense that during the day, my solar array produces more electricity than my homestead consumes. Because I'm tied to the grid, I can continue drawing power through the night, even though I'm not actively producing it. The same is true during the winter, when daylight hours are shorter. I still don't have a bill for electricity consumption, because during peak production, my solar panels produced enough energy to offset my off-hours and off-season use.

It's a satisfying moment to see your electric meter spin backward.

Net Metering Alternatives

While net metering is the most common way residential-solar producers are compensated for energy production, there are other ways. Some of these work out better for the consumer, and some work out better for the utility.

Net billing is like net metering, but the net energy consumption is calculated and canceled out each month. Energy credits accrued in the sun-abundant summer months would expire before the lean winter months arrived.

Solar renewable energy certificates (SREC) focus on the non-energy attributes of solar energy, like the environmental impact. Each megawatt-hour of electricity generated from a solar energy system is worth one SREC. Here in the United States, states encourage solar generation by requiring electric utilities to purchase a certain amount of SRECs. You may have heard this called a "solar carve out." With SRECs, you're not paid for the cost of your energy, rather you're paid for the values associated with renewable energy production. Only seven states and Washington, DC, have SRECs as of this writing.

The Value of Distributed Energy Resources (VDER) Value Stack Tariff is New York's version of net metering. It pays for electricity based on energy cost as well as other values—hence the value stack—like location, time of day, how much your energy contribution reduces the peak load on the utility's system, and more. This tariff is more complex than other energy compensation.

Here's an exaggerated example to illustrate net metering: If I've produced enough energy from my solar panels to equal $500 USD in electricity, my account will show a -$500 balance. Come December, when my energy production is low, I might consume $200 in electricity from the grid. The energy utility will take that $200 out of my negative balance, charge me $0, and update my balance to -$300.

Where things get messy is in rising electricity rates. In summer 2022, I racked up a great energy credit, but a hike in rates from my electric utility that fall disproportionately used up that credit. I ended up generating electricity at a rate lower than I later had to purchase it.

EPIC TIP

GO SMALL

Full-homestead solar arrays are the ideal for self-sufficiency, but if a project of that scale is out of reach or just too overwhelming, there are many ways you can harness solar energy on a smaller scale. For chickens, electric net fence and automatic coop doors can be charged with a simple, small solar panel and battery. In many parts of the world, solar water heaters are mounted to rooftops and used to heat water with great success, and these can be connected to electric or gas water heaters so you never run out of hot water in the shower.

Choosing Your System

Solar arrays are becoming both less costly and more available than ever before. I sound like one of those door-to-door solar salespeople—I still get visits from them twice a month!—but it's true. This is good news because you have the option to be picky about your system and your installer. Here are some things to look for in your solar research:

Price per watt. When making decisions, my friend and sustainable-tech expert Ben Sullins—Ben Sullins Official on Instagram and YouTube—stresses the price per watt generated.

The median cost of residential solar installation has fallen by an average of $0.40 USD per watt per year, over the time the Berkeley Lab has been keeping track. In 2021, these prices ranged from $3 to $5 USD per watt. The cost varies depending on where you live and what's happening in the supply chain when you make the investment.

Enjoying a well-lit dinner under the back awning, courtesy of solar power.

EPIC TIP

LOOK FOR REBATES

California in 2020, there was a 26 percent federal tax credit on any expense involved with owning and installing solar panels. This included roof work necessary for solar-panel installation. In the end, I was able to deduct $6,400 USD from my federal income tax.

You can find tax credits, rebates, and even grants for energy-related projects, from a new solar installation to improving energy efficiency in your home. Here are a few resources for rebates, incentives, and financing in the United States:

- Database of U.S. State Incentives for Renewables and Efficiency, by the North Carolina Clean Energy Technology Center at North Carolina State University: www.dsireusa.org
- Energy Star's guide to Energy Efficient Mortgages: www.energystar.gov/newhomes/mortgage_lending_programs
- Energy Star's list of U.S. federal income tax credits and incentives for energy efficiency: www.energystar.gov/about/federal_tax_credits
- Energy Star's rebate finder: www.energystar.gov/rebate-finder
- EnergySage has a great primer on the U.S. federal income tax credit at https://news.energysage.com/solar-tax-credit-explained

Warranties. Photovoltaic panels get less efficient over time but tend to not "go bad." My Panasonic panels are rated for a twenty five–year warranty. In twenty years, I'll probably lose 15 percent of my annual energy production. The panels aren't broken, they just become less efficient over time. You may be offered a performance warranty, which guarantees a percentage of energy production over time—for example, 90 percent of initial electricity production at ten years and 80 percent at twenty years.

Some components, like the power inverter, should carry their own warranty for ten to twelve years.

In twenty-five years, who knows what type of advanced energy systems will be available. At that point, once all of the warranties are up, it might make sense to get a new system. By this time, most setups have paid for themselves roughly two times over. I really hope we can recycle the components in these panels by that time, too.

Reputation. Going with a more reputable company, even if it means paying more, can save frustration. Use a company that's likely to be around in the future. If your installer closes up shop, you could have trouble finding someone to service or make improvements to a system they didn't install.

AC & DC: AN ENERGY LESSON

There are two types of electricity: direct current (DC) and alternating current (AC). Solar energy is collected as DC, but your home is run on AC. This is why you can't send solar power directly into your home's electrical system. As the photovoltaic panels collect energy, the energy filters through an inverter or a microinverter to convert DC to AC. From here, it's either directed into the electric grid or to your storage battery.

EPIC TIP

THREE PLANNING RESOURCES
Planning for solar energy requires a lot of estimating. These resources are useful for planning a system of your own:

- National Renewable Energy Laboratory's PVWatts Calculator determines how much solar energy you can expect to collect from your location: https://pvwatts.nrel.gov.
- Energy Sage finds quotes on solar arrays and installation in your area: www.energysage.com.
- Google Project Sunroof looks at the solar energy potential of your location; then factors in electricity costs, average solar installation costs, and rebates and incentives; and offers an estimate of what you can expect to pay overall: https://sunroof.withgoogle.com.

BATTERY POWER

If my power went out, I wouldn't have electricity because I'm tied to the grid rather than set up with batteries. The grid itself acts as my battery, absorbing my energy-production overage. If I had batteries, I would have a place to store the energy I'm producing, and I would have energy even in a power outage.

Adding battery storage to your solar array can bump up the price by $0.60 to $1.60 USD per watt, according to the Berkeley Lab (at 2021 prices). While it hasn't been on my priority projects list, a battery system is coming to the Epic Homestead soon.

Going Off Grid

One of the biggest reasons to get a battery is to give you off-grid protection for your energy needs. It's a good emergency-preparedness purchase. I live in the wildfire belt, and sometimes electricity is cut for fire mitigation. Rolling brownouts happen across the United States and elsewhere when energy demand is too high during heatwaves and unusually cold spells. There are plenty of opportunities for power to go out wherever you live, and those opportunities will only become more frequent. If you want to have an off-grid homestead, wherever you're located, you need a bank of batteries to remove yourself from the main power supply. Batteries are holding tanks for the energy you collect. When the refrigerator kicks on or you flip on a light switch, the energy leaves the battery, runs into a subpanel, and goes through your existing wiring to the appliance.

Rate Arbitrage

Another reason to get a battery for your solar array is to take advantage of the best energy rates from the electric company. Back to Ben Sullins: He uses his solar array and battery system for rate arbitrage—also known as peak energy shaving. He's essentially playing the energy markets, like an energy trader.

Here in San Diego, the electric utility uses time-of-use energy billing. Peak energy hours—the time when everyone is home from work and taking part in energy-intensive activities—are 4 to 9 pm. Energy rates are very high during peak hours, whereas at 2 am, they're very low. Rate arbitrage makes the most sense in areas that have a high peak-hour rate and a low off-hour rate, which we do.

Knowing your electricity rate plan and schedule is key to maximizing solar.

Ben runs his house off his battery during peak hours while his solar panels are sending energy into the grid. The energy he's selling back to the electric utility during peak hours is worth peak-hour rates. During the low hours, he uses energy from his solar panels to recharge his batteries. If he weren't already collecting all the solar energy he needs, Ben could use the grid to charge his battery, charge his cars, and schedule his laundry to run during off-peak hours and still come out with an energy credit. This doesn't just reduce his energy costs—it also reduces the stress on the grid by taking his home off grid during times of overall peak energy demand.

If you want to put the work into scheduling your homestead activity using a rate arbitrage model, this is a quick way to make the batteries pay for themselves. It might be next up on my energy-project list.

MINI-SPLIT HEATING AND COOLING

Being smart about energy use is a hallmark of sustainable living. The biggest energy drain in a home is heating and cooling—as much as 51 percent of your energy use, according to the University of Colorado Boulder Environmental Center. Stand by your electric meter when the AC kicks on, and you can watch the number of kilowatts used climb.

Heat pumps move heat from outdoors to indoors in winter and from indoors to outdoors in summer. These are significantly more energy efficient than other kinds of electric heating and cooling and may be eligible for energy-efficiency rebates or tax incentives.

A mini-split heat pump acts as a heater, air conditioner, dehumidifier, and fan, all in one unit. In each room, a rectangular air handler box is mounted at the top of the wall. Outside your home is a large unit, not unlike a central air-conditioning condenser unit. A conduit between the air handlers and the condenser houses tubes and cables. The condensation drains out a pipe on outside your home.

I was surprised to learn that more than 30 percent of energy in climate control is lost as the air moves through ductwork, especially in poorly insulated spaces. Mini-split heat pumps don't have ducts, so you don't have that loss.

In a small home and moderate climate like mine, a mini split made so much sense. My four units cost around $12,600 USD installed in 2020. A system with air handlers in only the most crucial rooms would've cost less. Central air, in case you're wondering, would've cost nearly twice that.

This outdoor unit pumps either hot or cold air into the home.

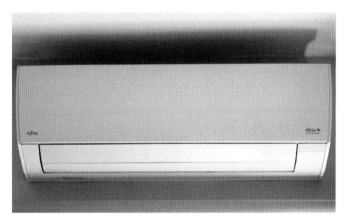
The inside units are sleek and can be tucked in the corner of a room unseen.

Mini-Split Technology

I love that I can set a different temperature for each room, turn the air handlers on and off as I need them, and program the air handlers to adjust the temperature based on the time of day. The mini split offers a lot of flexibility by zone, so I don't have to heat or cool a room I'm not using.

The wall-mounted units are remote controlled, and I connected them to a couple of apps for full automation. While my mini splits weren't app-ready out of the box, I set up the Cielo Breez Eco device and app to work my climate control. There are other similar apps out there, and you can find systems that already have connectivity. A good

app will allow you to adjust temperatures in each room, set timers, check the filter, and more.

I took the automation one step further and connected Cielo Breez Eco to my Google Home app. Now I can control the mini-split units with my Google Home Routines. Google Home also uses geofencing, so if it doesn't detect my phone on the property for a certain amount of time, it'll turn off the air handlers. All this technology ensures I'm putting energy into climate control only when I need it.

Mini-Split Installation

Unless you already have experience with HVAC systems, I recommend hiring someone to install your mini split. My unit's installation turned out to be more complicated than expected. The installers had to route the electrical through the crawl space under the house. The whole install took them two days, and if I'd tried to do this myself, it would have taken at least ten and I'd have holes drilled all over my house with no idea what to do next.

OUTDOOR LIGHTING

A homestead is a lot of work, and I want to enjoy this place as much as possible. The right outdoor lighting allows me to take advantage of my space at all hours. Too much outdoor lighting, or the wrong kind, is wasteful. Light pollution is harmful to birds, insects, and night-dwelling creatures, and researchers at the Iowa State University found it can even mess with plant growth.

Here are three environmentally thoughtful tips from the International Dark-Sky Association, plus one of my own, for your outdoor lighting:

- Shield fixtures so light is directed down, where it's needed, not into the sky and surrounding areas.

- Turn off outdoor lights when you don't need them. You might argue for outdoor lights as a general safety measure, but there's no actual evidence that lighting reduces crime.

- Use warmer color lights with a maximum of 3000 Kelvin. Blue light brightens the night sky more than warm light.

- Take advantage of timers. I have lights on a photosensitive timer to come on at dusk, and they stay on for about four hours, which is the amount of time I'd typically stay outside.

While installing a whole new climate-control system is a great energy saver, a less ambitious project with impact would be to upgrade your lighting. In Chapter 3, you read about the most efficient lighting for indoor food production. Like a plant, you'll enjoy your time indoors more with the right lighting.

LEDs

In the spectrum of energy-efficient lighting, incandescent bulbs are at the bottom of the list, with compact fluorescent lights (CFL) coming in second, and light-emitting diodes (LED) at the top. The U.S. Department of Energy (DOE) says LEDs use at least 75 percent less energy than incandescent bulbs. They can last twenty-five times longer than incandescents and four times longer than CFL bulbs. LEDs are also dimmable, and they're cool to the touch.

In 2010, the DOE came out with incredible statistics, estimating that if LED lights were rapidly adopted by 2027, we would save about $265 billion USD in energy costs, avoid the need to build forty new power plants, and reduce electricity demand for lighting by 33 percent. The cost of LED bulbs has dropped significantly since they were introduced, which is another attractive reason to switch.

Translating watts to lumens. If you grew up with incandescent bulbs or you have older light fixtures and lamps, you're used to choosing a bulb by its number of watts, which measures the power needed to operate the bulb. In choosing LED bulbs, you're instead looking at lumens, which measures the brightness of the bulb. If you want to replace a 75-watt incandescent light bulb, you'll be in the market for a 1,100 lumen LED. That LED bulb will use just 16.5 watts, which is a great illustration of your energy savings.

Soft light. When CFLs were the thing, a lot of us complained about the cool, blue light. We were used to the incandescents' warmer, rosier glow. LEDs come in a range of light colors. If you want a cooler, bluer light, you're looking for something with a high K-value (Kelvin color temperature); warmer, yellower light is found in the low K-value bulbs.

Hardwired, low-voltage LED lights use little power and add much-needed nighttime ambience.

Timers

As a kid, you were probably repeatedly reminded to turn off lights when you left a room. To kids who don't understand what it means to pay electric bills, wasting electricity means nothing. Now that you know better, turning off lights when you don't need them is second nature. Sometimes, though, you're not home when a light needs to turn on—like your porch light before you get home from work—or turn off—like your grow lights after they've run for sixteen hours. Enter timers.

Photosensitive timers. Photosensors prevent lights from operating during daylight hours. This is great for outdoor lighting. Many indoor nightlights have the same feature. The sensor uses ambient light conditions, switching themselves on and off as the light changes.

Smart home timers. I'm huge on these. In this chapter's section on mini splits, you read about integrating climate control into your whole-home app. Lighting works the same way. You can also use smart home hubs that route many switches to one hub, so you don't have 200 things connected to your WiFi network. Program Google Home—and others like it—to flip lights on and off and set up routines so your lights are on when you need them.

Analog and digital programmable timers. Without needing WiFi or whole-home connectivity to operate lights, an analog or digital timer plugs into the wall socket, and the light plugs into it. These are the original smart home timers! Before selecting a programmable timer, read the label to be sure it'll work with the light fixture and whether it's for indoor or outdoor use.

It may have been a while since you've seen the analog-style timers, but they're still out there—my mom puts out what seems like 100 for Christmas every year—and they still work. These timers have a dial with buttons representing every half hour in a 24-hour period. You depress the buttons for the time that you want the light to be on.

This fire pit sits over an underground pond water reservoir, so I opted for a sturdy metal model.

FIRE PITS

Back to the idea of wanting to enjoy the Epic Homestead as much as possible, a fire pit is a great way to do that. If you have a yard or even a patio and can swing a fire pit, I recommend one. Besides the entertainment factor, a fire pit reduces my need for outdoor lighting, allows me to dispose of dried and diseased plant clippings, and produces a valuable compost resource—ash.

Types of Fire Pits

For the purpose of fire pits as a homestead energy resource, I'm focusing on wood-burning fire pits here. The natural gas, propane, and gel-fuel fire pits are nice, but I don't think they add to sustainability.

Within wood-burning fire pits, there are a few types:

- **Traditional campfire-style fire pit.** Inexpensive to construct and straightforward to use, a campfire-style fire pit needs some kind of ring of bricks or stones to keep the embers somewhat contained. Two downsides to a campfire-style pit are that it's not portable and it poses more of a fire safety risk than other types.

- **Fire pit bowl.** A favorite for small outdoor spaces, they're easy to put away when you're not using them. The size of fire-pit bowls limits you in the size of fire you can have. This may be a good thing, if you tend to get carried away with your fires, or it could be a bad thing, if you're trying to burn a lot of garden materials.

- **Chiminea.** Attractive chiminea options seem to be everywhere now. These are great for containing fires, acting more like fireplaces than pits. They can be moved out of the way when not in use but are generally heavier than fire-pit bowls.

- **Barrel-type fire pit.** These fire pit styles are fun. Newer styles have well-spaced ventilation holes that work with the flames. They burn very efficiently, and you may end up with a gigantic fire, if you're not used to using one. They're smokeless, and the tall sides keep flaming materials in the pit, where they belong.

Fire Pit Safety

Why is someone living in an area prone to wildfires offering fire pits as a homestead energy resource? Because there are ways to safely have a fire in your backyard. This is assuming you can legally have a fire pit to begin with. Know your city and county ordinances and burning rules.

To start, fire pits are tools for responsible adults only. I don't mean to be a buzz kill here, but fires take on a life of their own and safety can't be stressed enough. Before starting the fire, know who's responsible for keeping kids and pets at a safe distance.

Have a fire extinguisher or hose with a ready water supply nearby, and don't walk away from the fire. Definitely be sure the fire is all the way out before you leave it for the night. Embers can stay hot for hours and sometimes days after flames are extinguished.

Locate your fire pit away from dry brush and hanging branches. If you're using a portable fire pit, put it on a level surface. Don't use a fire pit in a screened area, and keep it at least 10 feet (3 m) away from structures and combustibles.

Understand what you're burning. Untreated wood is good; chemically treated wood is bad. Paper is handy for getting a fire started but can get caught on a breeze while it's alight. Lighter fluid—while it seems like a good idea at the time—causes fires to get too big, too fast. Make your own fire starters instead, with dryer lint or egg cartons. While some garden scraps are okay to burn, never burn poison ivy, poison oak, or poison sumac, because urushiol particles are carried in the smoke and can cause severe respiratory problems.

Using Fire Pit Ashes

The wood you're burning in the fire pit doesn't just disappear in the fire. It's reduced to ash, which is a carbon-rich organic material. Used in moderation, ash can improve compost.

Ash is high in calcium, potassium, and salt, and it has a very high pH. There are a lot of nutrients in ash, too, in lesser amounts: magnesium, phosphorus, sulfur, and more. The exact amounts vary by the type of wood and whether the fire was burning hot or cool.

These elements are in delicate balance in the garden already, so use your soil test to see if ash is a good addition to your compost. Don't add ashes to your compost if your soil has a pH of 7 or higher, is high in potassium, or has high salinity. Before you take a few shovelfuls of ash to the compost pile, think about what you burned in the fire pit. If there was anything inorganic or potentially toxic, don't use those ashes. Likewise, if you're burning wood from trees grown in potentially contaminated soils, this wood could be carrying heavy metals and other toxins. Be careful that you don't breathe in ash particles or get them in your eyes and on your skin while you're transferring ash from the fire pit to the compost bin. Wait until those ashes are totally cool before moving them, too, both for your safety and the safety of your compost microorganisms.

Now that we've talked about integrating smart energy systems on the homestead, it's time to dive into some useful ways we can conserve another important resource—water.

Checking the 5,000 gallon (19,000 L) rainwater cistern levels. This system provides over a month's worth of water to my garden during dry spells.

chapter seven

WATER CONSERVATION

FROM FAR TOO LITTLE RAIN TO FAR TOO MUCH RAIN, there are few places with "normal" amounts of water resources anymore. Even in an area like mine—known to be arid—weather extremes are turning upside down what we thought we knew about water.

Harvesting and conserving water resources are becoming more important as rainfall becomes more uncertain. One year you may have it, and the next you won't. I put in place a number of water-wise systems and practices at the Epic Homestead, and there's even more that can be done on homesteads large and small.

USING GREYWATER

When fresh water runs through your laundry or shower, what you're left with is greywater. This slightly used but relatively clean water can be reclaimed and put to work elsewhere. It serves two purposes: You lighten the strain on your municipal sewer system or home septic system, and you reduce the amount of fresh water your homestead needs to operate. If everyone diverted their shower and laundry from the sewer into a greywater system, we'd cut the sewage treatment burden by about 50 percent.

Plumbing your home for greywater isn't legal everywhere, but on the other hand, some cities have rebates to encourage you to install an approved greywater system. Check your building and plumbing codes.

Scan for video

The shower greywater system supports my entire front yard orchard with safe supplemental water.

Designing Your System

Greywater output. Have an idea of the amount of water you expect to run through your greywater system each year.

Washing machines can use 7 to 35 gallons (26-132 L) of water per load. Your machine's specs will tell you how many gallons each load runs through. Multiply that by the approximate number of loads you do in a year.

The U.S. Environmental Protection Agency estimates a standard showerhead uses 2.5 gallons (9.4 L) of water per minute. Shower heads with a WaterSense label use 2 gallons (7.6 L) or less. Figure out what your shower runs per minute, multiply that by the length of your shower, and multiply that by the approximate number of showers taken in your home in a year.

With these numbers, you can determine how much water will flow through a greywater system on your homestead.

Greywater for gardens. Look at the area you have for greywater distribution, and also look at what you're trying to grow there. Greywater can be used to irrigate trees, shrubs, ground cover, and lawns. Don't use greywater where you're growing root crops or fruits and vegetables that touch the ground.

My laundry system puts out 500 gallons (1,892 L) per year. The laundry greywater runs to a garden area just behind the house, where I put artichokes. Artichokes are short-lived perennials that like frequent watering. They grow well above the ground, avoiding potential contamination, so they were an easy match for this system. When the artichokes die back, I plant flowers to soak up that juicy greywater until the 'chokes come back.

At the same time, I sent greywater from both my bathroom shower and my outdoor shower across the yard to the citrus grove. I planted saplings here and actually ran the risk of overwatering my citrus with my showering. I monitored them closely, and it turned out fine. I'm considering extending these basins to be able to add as many as eight more trees.

Laundry greywater is enough to keep perennial edibles alive without pulling from the city tap.

This risk of overwatering is one reason a diverter valve is a crucial part of greywater design. During a period of heavy rain or heavy greywater use—and when you notice signs of overwatering, like plant stress and saturated soil—you can turn a valve and send the greywater into the sewer or septic system, instead of to your plants. I made the mistake of not installing a diverter on my outdoor shower, and I wish I would have.

Mulch basins. Greywater shouldn't run from the system as surface water. A mulch basin is the route I chose to get the water to my plants. This is just what it sounds like: a basin—or trench—dug out and filled with mulch, used to flood-irrigate areas of the garden.

The mulch acts as a wick. The water drains out of the pipes from the shower and laundry, into the bottom of the basin, which is just soil. Some of the water drains from the basin right away, and some of it pools. The mulch wicks up that water and spreads it around the entire basin. As the soil around it needs water, it draws from the mulch.

An abundance of artichokes solely watered with laundry greywater.

A mature orchard can't live solely off greywater, but using it significantly reduces water costs.

The irrigation set-up of the laundry greywater system before adding mulch into the basin.

Basic maintenance. Installing a greywater system is an undertaking, but once the system is installed, maintaining it is straightforward. There are only a few things you need to do:

- **Keep emitters clear.** Irrigation boxes house the emitters and keep them safe from debris. Eventually, the mulch breaks down, and the irrigation boxes may need to be cleared of the fines. Open the lid and look inside once a month or so.

- **Add more mulch.** Because the mulch breaks down over time, it'll need to be topped up. How often depends on a lot of things, including the type and age of the mulch, the amount of water flowing through the basin, and your weather.

- **Flush salt buildup.** Clay soils add another maintenance item to the list. Salt buildup is possible here because the salt contained in greywater doesn't filter through heavy soil as well. When you check the irrigation boxes, also clear away some of the mulch and look at the soil in the basin. If you see salt crystals, rinse them out with fresh water. As the mulch breaks down, it'll naturally improve the soil's water-draining capacity around the basin and reduce salt retention.

- **Reshape the basin.** After a year or so, you'll notice some natural misshaping as water and weather move soil. If needed, plan to touch up the basin and even reshape it for your own reasons now and then.

Laundry Greywater

Greywater works well in a laundry system partly because washing machines already have a built-in pump, and that pump—combined with gravity—will push out the water to your irrigation field.

Laundry greywater requires just one habit change, and that's to use a biodegradable detergent, which you might be using already. I use Oasis Laundry Detergent. It's hyperconcentrated, and as of this writing, the label claims the product breaks down into carbon dioxide, water, potassium, and sulfur. It avoids the phosphate, sodium, and plant toxins we don't want in our soil, and it won't raise the pH. Other biodegradable detergents can be appropriate, too. Always check the labels.

Irrigation valve boxes protect your water outlets from damage, clogging, and degradation.

Oasis laundry detergent is my go-to choice for food-safe laundry water.

INSTALL A LAUNDRY GREYWATER SYSTEM

Digging a thin channel for the PVC irrigation pipe carrying the laundry water.

Materials

Vary according to your specific project's size and setup.

I'm not going to lie: Installing a laundry greywater system is not the easiest homestead project. While Jacques Lyakov and I did the exterior work, I enlisted a local service provider, CatchingH2O, to plumb the system at The Epic Homestead. You might have more home-construction chops than I do, and in that case, this is something you could handle with the right set of plans. The project here is just an outline of steps, not meant to be a step-by-step guide to your own setup.

An invaluable resource for laundry greywater installation is the Laundry to Landscape open-source concept by Oasis Design: https://oasisdesign.net/greywater/laundry.

Building Instructions

1. Dig your trenches. Decide where you'd like your greywater to run. Dig two trenches: One trench to get the pipe from your house to the plants, and one to route the water throughout the planting area—this one is the mulch basin. The pipe trench only needs to be a few inches across. Make the mulch basin about 18 inches (46 cm) across.

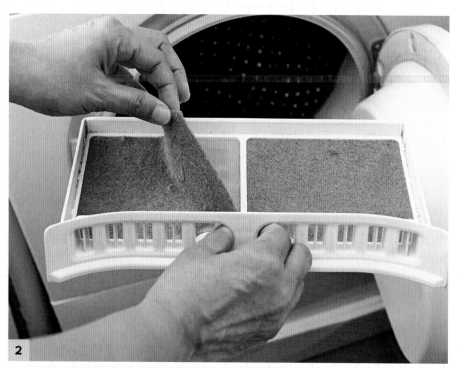

If you're worried about microplastics, you can install a filter to capture them.

To use gravity in a flood-irrigation system like this, you need to drop the flow at least ⅛ inch per foot (3.2 mm per 0.3 m). The point farthest from the water source is the lowest point in the system.

If you live in an area that freezes, you either need to dig a trench deep enough to get your pipes out of the freeze zone, or you need to use the diverter to send your greywater into the sewer or septic system during freezing weather. If you're DIYing this system, using the diverter is the better option—there's not as much to water in the garden in the dead of winter, anyway.

2. Install a microfiber filter. Our clothing sheds microfibers in the wash. From synthetic materials, like spandex and polyester, those fibers are actually tiny bits of plastic. A filter isn't essential, but it will keep microfibers from clogging the irrigation emitters and contaminating your soil.

3. Install a diverter. Using your washer's existing plumbing, install a three-way valve that will allow you to send water into the sewer/septic system or into the greywater system. Put this valve in a place that's easy to reach, and label it.

4. Install a pipe to the sewer/septic. From one side of the diverter valve, run a pipe to your existing sewer or septic drain.

3

This actuator toggles the shower water to go to the sewer or to my orchard with the flip of a switch.

4

This actuator connects to standard plumbing fittings.

5

Drilling a hole in the side of the house to thread the irrigation pipe through.

7

Inside the irrigation cover, a valve can be opened or closed.

9

Water should seep to over twice the width of the irrigation basin.

5. Install a pipe to the outdoors. Drill a hole through the laundry room wall to the exterior of the house. I lucked out by having the washing machine against an outside wall. If yours is in an interior space, you'll need a more complicated plumbing connection to route the water out of the house. From the other side of the diverter valve, run a pipe through the hole you drilled in the wall.

6. Install an auto vent. Now outdoors, install a T at the end of the pipe that's exiting the house. Running from the top of the T, install an auto vent.

7. Install the pipe to the landscape tubing. From the bottom side of the T, install a pipe that connects to the tubing that will run through the mulch basin.

8. Install the landscape tubing, emitters, and irrigation-control valves. The tubing that runs through the mulch basin, the emitters spaced throughout, and the control valves come next. Leave the end of the tubing open so the system drains properly.

9. Install irrigation valve boxes. Rigid plastic boxes around the emitters and valves protect these parts from getting clogged or crushed.

10. Test the system. Run a load of laundry to be sure the water is flowing properly.

11. Fill in the mulch basin. Leave the irrigation valve boxes visible, or somehow mark their location so you can find them easily when needed.

Shower Greywater

Shower greywater works a little differently than laundry greywater in that a shower doesn't have a pump to move water out of the system. You're relying only on gravity to move water. When it comes time to dig trenches to move the water through the pipes, you absolutely want to be sure you're getting it right. Stagnant water sitting in a pipe will create an odor and a potentially bad bacteria situation. Water could also run backwards through the system if the gravity flow is off.

Like in a laundry greywater setup, be careful which soap products you're using. This could be a big adjustment if you're particular about your hair care and body-care products. Say goodbye to silicone and other nonbiodegradable ingredients! Water-conservation consultant Brooke Sarson, of CatchingH_2O, recommended to me Aubrey Organics, Avalon Shampoo, some products from Burt's Bees, and Dr. Bronner's—my personal choice. Be sure to read the label.

HOW **NOT** TO USE GREYWATER

For all of its practicality, there are a few things to avoid in using greywater.

- Don't include your kitchen sink and dishwasher in a greywater system. The oils and food particles coming from your kitchen can clog emitters, cause bacterial growth, and throw the whole system out of balance.

- Toilets are considered blackwater, not greywater. Blackwater is too contaminated to grow anything, even ornamentals.

- Don't send your greywater into an area where people and pets will interact with it. Keep it off of your lawn, and definitely don't run it into a public area.

- Don't try to store greywater. It's not fresh water and will go foul.

- Don't use greywater in drip irrigation. The particles that invariably end up in greywater could clog emitters and sprayers.

- Don't spray greywater airborne. It's against the code—in California, at least.

- Don't use soaps and detergents with salts, boron, or chlorine bleach. Bar soaps, too, can change the pH of the water.

USE AN OUTDOOR SINK

The simplest greywater install of all may be an outdoor kitchen sink like mine. I use this sink to wash and do basic prep for vegetables before bringing them into the kitchen.

Here, I wash off dirt and debris from my harvest, then collect that water in a bucket and use it right away to water whatever needs it. This option both keeps my kitchen cleaner and allows me another water-reuse opportunity.

A word of caution: Don't leave a bucket with water in it sitting around. Birds and wildlife can drown in it, and stagnant water becomes a breeding ground for mosquitoes.

As long as it doesn't include meats, fats, or anything harmful, you can capture your sink water and use it to water the garden.

RAINWATER CAPTURE

Rainwater collection at the Epic Homestead started out looking something like a Rube Goldberg machine. I hastily pieced together flex tubes, IBC totes, and plastic barrels to catch a few big rains, which they did—but it was messy, and they didn't catch as much water as they could have. I knew I needed a better plan, so I looked for grant funding to expand the project, worked with water-conservation consultant Brook Sarson, and now have a great system.

Between three cisterns at the Epic Homestead, I have a total rainwater storage capacity of 5,700 gallons (25,912 L). I use about 80,000 gallons (363,687 L) of water a year for the whole homestead, including all home use like showering, cooking, etc. Considering this, what I'm able to catch and store is not some amazing amount, but we don't get a lot of rain in San Diego—just 10 inches (25 cm) a year on average. I want to harvest and use as much as I can.

As a rule of thumb, you can harvest 600 gallons of water per inch (2,271 L per 2.5 cm) of rain from a 1,000-square-foot (93 m²) roof. Depending on your roof's shape, the wind, and other conditions, this amount could vary. There's a formula to help you get a more exact number:

Gallons of water harvested = square foot of roof catchment area x inches of rain x 0.623 conversion factor (liters of water harvested = m² roof catchment area x mm of rain).

See the Rooftop Water-Catchment Calculator on page 158 to look at how much water you can collect from your roof in a year. This number may help you decide whether rainwater capture is right for your situation and, if it is, how much holding capacity you want to install.

My first rainwater capture system was a bit makeshift, but it got the job done.

ROOFTOP WATER-CATCHMENT CALCULATOR

WATER COLLECTED OF ANNUAL RAIN IN INCHES (CM)	SIZE OF ROOF IN SQUARE FEET (M²)					
	800 (74 m²)	**1,000** (93 m²)	**1,500** (139 m²)	**2,000** (186 m²)	**2,500** (232 m²)	**3,000** (279 m²)
10" (25 cm) rain	**4,984 g** (22,658 L)	**6,230 g** (28,322 L)	**9,345 g** (42,483 L)	**12,460 g** (56,644 L)	**15,575 g** (70,805 L)	**18,690 g** (84,966 L)
20" (51 cm) rain	**9,968 g** (45,315 L)	**12,460 g** (56,644 L)	**18,690 g** (84,966 L)	**24,920 g** (113,288 L)	**31,150 g** (141,611 L)	**37,380 g** (169,933 L)
30" (76 cm) rain	**14,952 g** (67,973 L)	**18,690 g** (84,966 L)	**28,035 g** (127,450 L)	**37,380 g** (169,933 L)	**46,725 g** (212,416 L)	**56,070 g** (254,899 L)
40" (102 cm) rain	**19,936 g** (90,631 L)	**24,920 g** (113,288 L)	**37,380 g** (169,933 L)	**49,840 g** (226,577 L)	**62,300 g** (283,222 L)	**74,760 g** (339,866 L)
50" (127 cm) rain	**24,920 g** (113,288 L)	**31,150 g** (141,611 L)	**46,725 g** (212,416 L)	**62,300 g** (283,222 L)	**77,875 g** (354,027 L)	**93,450 g** (424,832 L)
60" (152 cm) rain	**29,904 g** (136,037 L)	**37,380 g** (169,933 L)	**56,070 g** (254,899 L)	**74,760 g** (339,866 L)	**93,450 g** (424,832 L)	**112,140 g** (509,799 L)

Rainwater Prep

One of the first projects I had done when I moved to the Epic Homestead was a roof replacement. I didn't get around to putting up new gutters right away, so seamless gutters were my first purchase for the rainwater-collection project. If you also need a new gutter installation, check with your city before beginning. Some have rebates.

Copper was my choice for gutter material. I'll admit this was the fancy option: There's nothing wrong with aluminum or steel gutters, but those don't look as nice or last as long. Copper was 2.5 times more expensive than aluminum and could be targeted for theft, but copper looks nice and will only get better as it ages and patinas. This is the kind of thing I splurge on, instead of cars and clothes. I ended up paying around $2,200 USD for the gutters installed, and they should last for the life of the property.

All the rainwater that falls on my entire house comes out of this one outlet—a challenging design problem!

The Rainwater System

The gutters collect the water off your roof. From here, it runs through a simple but ingenious system: The gutter drains into the leaf filter, which keeps leaves and debris out of your collection system. Next is a first-flush filter. This pipe collects the first 10 gallons (45 L) of water in a storm and diverts it away from the collection system. That's the dirtiest water, having just washed your roof clean. When that pipe is full, the remaining rainfall goes to the rainwater collection pipe.

Rain transfers from the gutter to the filtration system.

The first flush filter can be drained to reset it for future rains.

A first flush filter captures the first 10 gallons (38 L).

A leaf filter captures large debris before it enters the system.

Using gravity, rainwater travels down this PVC pipe and into the cistern.

In my system, the gutter on the shed releases water through the filters and into a rain barrel. From my house, the gutters connect to a rainwater pipe that runs underground to a feeder pipe on the 5,000-gallon (22,730 L) cistern. The entry point to the cistern is lower than the gutters on the house, so water flows right into the cistern as the path of least resistance.

I can run water from the valve on the rain barrel through a hose to refill the pond, soak the banana trees, and do longer waterings on other fruit trees and water-hungry crops. When it's time to use the water from my cistern, I open a valve at the bottom, and gravity lets the water flow out through a pipe. On a sloped property, it's possible that's all I'd need to be able to use this water effectively. I don't have a sloped property, so I connected a pressure-sensitive pump to the outgoing pipe. The pump gives the water hose-level pressure that'll reach anywhere on the homestead.

I use captured rainwater to irrigate my pond landscape as well as fill the pond itself.

IRRIGATION

Whether you have a wet climate or a dry one and are homesteading with a container garden or in-ground beds, watering plants is a regular chore. You can do this by hand with a hose or watering can, or you can set up irrigation.

Even before I had a garden as large as what's at the Epic Homestead now, I liked having an irrigation system for a few reasons:

- Targeted irrigation makes sure only the areas that need to be watered get watered. It saves a ton of water: 30 to 70 percent compared to sprinkler irrigation. It could save even more compared to hose watering, depending on how heavy you are with a hose.

- Irrigation set to run on timers takes one thing off my to-do list.

- I can run irrigation when it makes the most sense for the plants—early in the morning or late in the evening—even if I'm not home.

- Today's irrigation technology covers all of my irrigation needs, from direct-seeded cover crops and vegetables in neat rows to container-planted fruit trees and herbs growing in the landscape.

Nothing saves me more time at the homestead than a properly installed drip irrigation system.

Direct Irrigation

Conserving water in a garden means using water as efficiently as possible. You want to deliver the water right to the plants. This is next-to impossible with lawn sprinklers and time consuming with hoses and watering cans. There are several efficient, direct irrigation options:

- Soaker hoses put out water consistently across the entire length of the hose, which is great for direct-sown plants in a row, like carrots, beets, dill, and lettuce mix.

- Drip tape is the choice of market farmers and makes sense for annual gardens with straight rows. At certain intervals—usually 6, 12, 18, or 24 inches (15, 30, 46, 61 cm)—a hole in the tape allows droplets of water to fall from the line. This delivers water right to the base of the plant—if you've planted accurately along that interval. If you're off by a few inches in your planting, the plants will still do fine. Drip tape might last you a couple of seasons, but it's not known for its longevity.

- Drip line makes the most sense for perennial plantings and container gardens. A drip line is like a small hose, and from there, you run an emitter to each individual plant.

- Micro sprinklers will water the whole surface of the bed. You can set these to spray over large or small distances. They're more efficient than the lawn sprinklers that deliver large drops of water in a way that's hard to control. Micro sprinklers will water a whole bed of cover crop or other densely seeded plants.

The installation for each of these is similar to start and then becomes more specific as you go. Coming off of your water source, these irrigation systems begin the same way: with a backflow valve to prevent garden water from making its way back into the source, a filter to catch any fibers or debris in the line, and a pressure reducer to keep the water flow at the right rate for the system.

Once the water reaches the bed, each bed will have a header line. This is a main line that all the water-delivery lines run from. At the end of each water-delivery line, you need an end cap to keep the water from running out the other side.

At the end of the main line, I also like to install an end piece with a valve. The water sitting in the black tubing gets hot in the summer, so I can open that valve for thirty seconds to flush the hottest water before irrigating.

In between all of these main components are simple fittings. The whole system's installation is easy enough that you can do it yourself following a few YouTube tutorials and the manufacturer's instructions.

Typical Layout of a Drip Irrigation System

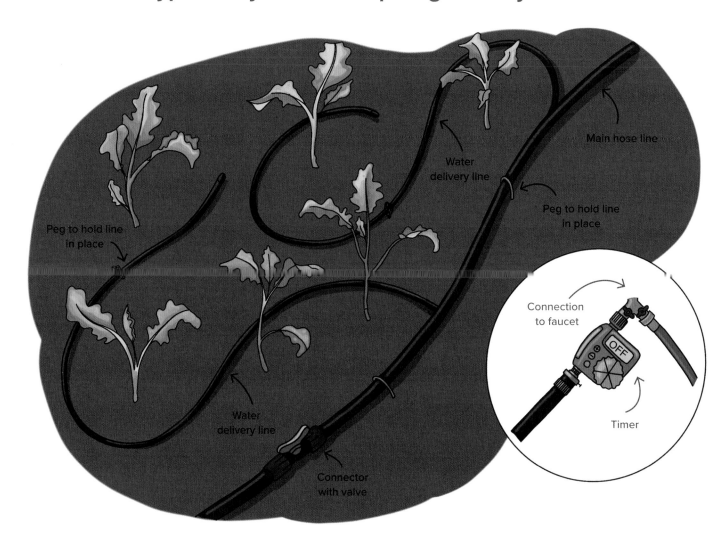

Main hose line

Water delivery line

Peg to hold line in place

Peg to hold line in place

Water delivery line

Connector with valve

Connection to faucet

Timer

Irrigation Zones

Setting up my 15-zone irrigation project has been the most worthwhile garden task I've completed to date. Each zone is on automatic timers, and this has cut down on time and excess watering in the garden. Changing a watering routine is now a matter of simply tweaking the system. For example, in the early rains of 2023, I was traveling for work and was able to turn off all fifteen zones remotely, so I wasn't wasting water and unnecessarily soaking my gardens.

Setting up zones is more complicated than running irrigation lines. You can set up your own zones and controller, though you might have a better overall experience working with someone who can do it for you.

Digital controllers. Multi-zone irrigation controllers come in all levels of complexity. The most tech-driven controllers are like smart home systems for your garden. You mount a digital panel to the side of your house or garden shed and install an app on your phone, and from either place, you can program or override watering routines for each zone. Slightly less advanced digital controllers have programming options at the panel without smartphone connections. Battery- and solar-powered controllers work for off-grid homesteads and gardens without electricity. Some controllers even have weather sensors that will hold off on watering when it's raining or freezing and will water more when it's windy to account for evaporation.

Manual zones. Without using programmable technology, you can manually set up your own zones using valves to turn the water on and off by hand.

One example of putting these valves to use is in a raised bed with three rows of drip tape. Along each of the outside rows, you might plant a slow-growing vegetable, like broccoli. Along the middle row, you could plant a fast-growing vegetable, like radishes. When you seed these plants, all of the rows need consistent watering. You'll harvest radishes in four or so weeks, and the broccoli will take more like eight weeks. After harvesting the radishes, turn off the valve for the middle row so the broccoli plants can continue to get the water they need without wasting water on the radish row that's already been harvested.

My drip irrigation manifold sends water to different locations around the homestead.

Higher-end models include a controller which can be managed via phone.

For further control, I install on/off valves on each drip line in a garden bed.

Passive Irrigation

Considering all the technology that we're using for irrigation now, it's easy to forget that irrigation itself is an ancient technology. Passive irrigation lets the soil absorb as much water as it needs when it needs it. Ollas are the simplest—and my favorite—means of passive irrigation. Ollas are thought to have originated more than 2,000 years ago. In arid areas, an unglazed clay pot, buried to its neck, is a tried-and-true method for making the best use of water. You can save 50 to 70 percent of water versus standard hose watering by using an olla.

Here's how it works: Bury the olla to its neck in a garden bed. The ideal olla design has a narrow neck, to prevent evaporation, and a bulbous bottom, to hold water. Take the lid off the olla, and fill the vessel with water. Let the soil wick the water from the olla into the plants' root zones. Keep the olla filled at least halfway all the time. Plant in a ring around the olla so the seedlings can get the water they need ASAP. In the olla video on the Epic Gardening YouTube channel, you'll find a handy chart of how far water will seep into the soil based on the size of the olla, as well as build-your-own olla instructions.

How to Use an Olla

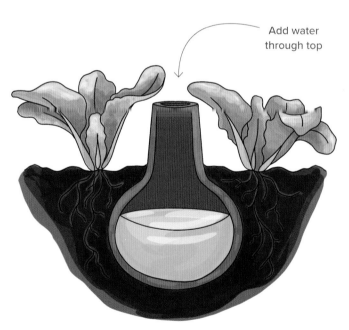

Add water through top

Water slowly permeates into the surrounding soil

You don't need water if your soil is still moist, which is exactly what mulch helps you achieve.

MULCH

You read a little about mulch in Chapter 2. Here's more info about how mulch plays into water conservation on the homestead.

There are some things we can predict from nature, and one of them is that nature will not allow the ground to sit uncovered for long. It's a brilliant system. But when we leave it up to nature to cover bare soil on its own, that ground cover isn't always what we, as homesteaders, want to see growing there—ahem, weeds and Bermudagrass. Bring in mulch to the rescue.

Why Mulch

Mulch is a great water-conservation strategy for a few reasons:

- It reduces the amount of water lost from the soil to evaporation caused by direct sunlight.
- It keeps the soil cooler, also reducing evaporation.
- It reduces weed growth, so the weeds don't take water away from the plants you've intentionally put in the ground.
- In the case of organic mulch, it breaks down over time, releasing nutrients into the soil and feeding soil life, which increases soil organic matter and improves the soil's capacity to hold water.

Straw is cheap and accessible.

Arborist-sourced woodchips are hard to beat.

If you have a lawn, repurpose your clippings.

A lush cover crop protects the soil.

Types of Mulch

There's a mulch for every purpose. The coarser the mulch material, the longer it will take to break down—which has both positive and negative sides—and you don't want it to be so lightweight that it blows away in a storm. Organic mulch is the ideal, especially around areas where you're growing food.

These are some mulch options I like on the homestead:

- **Straw.** Bales of straw will do for large areas, and bags of chopped straw are ideal for raised beds and container gardens. Ask for straw that hasn't been treated with herbicides because those chemicals can keep your plants from growing. Know the difference between straw and hay. Hay is more likely to have seeds in it, and then you're inviting more weeds to your garden.

- **Wood chips.** Use a chipper-shredder to break down woody material from your own lot, or bring in wood chips from elsewhere. Check out Use Free Wood Chips on page 47 for an Epic Tip on sourcing.

- **Pine needles.** Used a few inches deep, pine needles don't mat together, and they break down relatively quickly. They're acidic but don't contribute to lowering the soil pH when used as a mulch.

- **Grass clippings.** Bag up clippings from your own lawn or a neighbor's, as long as they're not sprayed with chemicals. Use fresh clippings at just ¼ to ½ inch (0.6 to 1.3 cm) or dry clippings at 2 to 4 inches (5 to 10 cm). The dry grass clippings are more likely to blow around.

- **Leaf mulch.** Shredded leaves will stay put better than whole leaves. Just run a mower over them before putting them in the garden.

- **Living mulch.** Also called green manure, living mulch is any ground cover that you grow between plants or as a cover crop.

Setting your homestead up for success in the water conservation department is a smart decision, from both a financial and a sustainability standpoint. As you can see, there are so many ways to do it. Pick and choose the systems that will work best on your homestead. Next, let's look at some mini livestock you can welcome onto your homestead without an excess of effort or expense.

SWALES

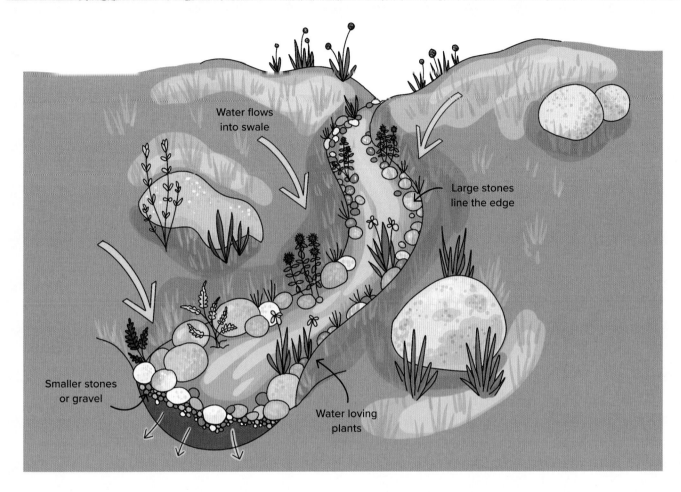

Water flows into swale

Large stones line the edge

Smaller stones or gravel

Water loving plants

Homesteaders with property have a great opportunity to direct water flow on the land so as much water as possible can be absorbed by the soil. Swales have been used for centuries as a coupling of simple earthen berms and trenches to move water along a path, following gravity and the slope of the land. Even on a property as small as the Epic Homestead, the swales in the backyard have made a difference in keeping rainwater from carrying sediment into the pond.

As water flows, the surface area and any vegetation planted in the swale slows the water and gives it a chance to soak into the ground. The longer and more winding the swale, the more water your land can absorb. Swales can look like typical landscaped areas, or yours might be lined with large stones. You could plant a rain garden or dig a small retention basin at the end of the swale to filter or capture the water.

Build your swale to take advantage of any natural slope you may have. Dig the trench at least 6 inches (15 cm) deep. Use the soil you dug out of the swale to build up the berm on the downhill side, which will catch even more water.

The University of California Master Gardeners of Butte County have good advice for locating swales in the yard:

- Keep swales at least 5 feet (1.5 m) away from homes or sheds without a basement; 10 feet (3 m) from homes with a basement.
- Don't build swales near a septic system or leach field.
- Locate trees' major root zones, and don't disturb the root zone with a swale.
- Be sure whatever you're planting along the swale has the appropriate amount of sun exposure.

Keeping chickens offers companionship, fertilizer, and, of course, eggs. Few animals are as productive on the homestead.

chapter eight

MINI "LIVESTOCK"

I'VE WANTED CHICKENS FOREVER. I'd never lived anywhere that would allow me to keep chickens until I moved to the Epic Homestead. Zoning restrictions, space limitations, and neighbors have made the livestock aspect of homesteading tricky—not to mention the time, skill, and money that goes into the actual care and feeding of these creatures.

As I've explored homesteading, I've come to realize that livestock doesn't just mean cows and sheep. On a modern homestead, mini livestock have so much to offer, from meat and eggs, to honey and pest control—and let's not forget the nutrient-rich compost contributions they leave behind. In this chapter, we'll look at chickens, ducks, quail, and bees.

CHICKEN KEEPING

The reasons to keep chickens are many. I like knowing the hens are kept with the best care and the eggs are hyper-local and high quality. I appreciate that the waste they leave behind is nutrient rich for the compost, and my quality of life is improved by having chickens around.

There are dozens of chicken breeds—some that are very good at laying eggs, some that are bred for producing meat, and some that do a little of each. I'm only talking about egg layers here but know that you have other options. Remember to check your local ordinances long before you purchase any equipment or birds to be sure chicken keeping is allowed in your community (see Permits and Zoning page 188), and to ensure you're following any legal codes that are in place.

I can't tell you how many chickens you should get. What I can say is that you should look at the amount of space you have to dedicate to chickens, the amount of money you want to spend on feeding them, and the number of eggs you hope to have, and come up with a reasonable number from there. I also should caution you about "chicken math," which is a multiplication program that proves you will always end up with more chickens than you intended. That is to say, it's hard to stop collecting them.

Scan for video

Trying (and failing) to capture Lobster, my Rhode Island Red.

Along those lines, a common misconception about keeping chickens is that they'll lay eggs every day. This is not quite true. The number of eggs you'll get depends on a few factors, including the breed, age, and health of the hen, and the time of year and your weather. In summer, my hens lay about one a day. In the winter, that falls off. As they get older, they'll lay fewer eggs. Realize that sometimes you'll be eating omelets every day, and sometimes you'll hardly have enough to make a cake.

To start your flock, you have the option of brooding eggs, bringing in day-old chicks, getting pullets—like chicken teenagers—that are a few months old and just about ready to lay, or finding adult hens. The level of chicken-keeping difficulty decreases as the chicken's age increases, but I'm convinced that raising your flock from baby chicks is the most satisfying experience.

Raising Chicks

I just couldn't help but order six-day-old chicks to get my flock started, so I got a crash course in raising baby birds. There's more of an investment here, because in addition to the coop, run, and everything else that adult birds need, you have to supply the chicks with essentially scaled-down versions of all of these things.

Here's the shopping list I used in preparation for welcoming my baby chicks:

- **Heat source.** Chicks can't regulate their own body temperature, so you need to provide them with the right amount of heat. Heat lamps have traditionally been used but are a fire hazard and increase the likelihood of pasty butt issues (a condition where poop gets caked around their hind end), so I opted instead for a radiant heater.

- **Feeder.** Chicks and chickens—birds of any kind— will make a mess of their food unless it's contained. A gravity-fed feeder costs just a few dollars and does the trick.

It's incredibly rewarding to raise your flock from baby chicks if you can swing it.

- **Waterer.** Like their food, chicks will be quick to soil their water, as well. A simple gravity-fed waterer comes to the rescue. I kept the waterer and the feeder raised off the ground to lessen the amount of bedding the chicks wanted to kick into them.

- **Bedding.** A bag of wood shavings or industrial hemp bedding will absorb moisture and odors, and that's all you need.

- **Brooder enclosure.** In addition to needing a place to stay warm, chicks need a place that's safe from predators, because they have absolutely no protection mechanism—apart from their cuteness. I put a stock tank in my dining room and used that as a brooder until the Epic Chickies were ready to move to the coop.

- **Feed.** Chicks also need age-appropriate food. Begin feeding a starter feed for a couple of months, then go to a grower feed as they develop their feathers and their growth really takes off. You'll eventually transition to a layer feed, which is designed to give them calcium and other nutrients for egg development.

The baby chicks quickly learned to use the brooder heater as a means of escape.

As your hens mature, they'll develop their own unique personalities.

Keeping Layers

Pullets are ready to move into their coop when they lose their fuzzy baby feathers and their adult feathers grow in. You still have a few weeks between moving the young birds to the coop and the time that you get eggs.

As adults, chickens require less maintenance than they did as chicks. They still need food, water, and protection, but they can now regulate their own body temperature. While they've lost some of their cuteness, their behavior is still entertaining, and—the main event—they're giving you eggs that you should collect every day. There are occasional things that can go wrong with your chickens, like having a sick or broody hen (one that won't get off the nest), and with regular interaction, you'll notice these issues pretty quickly.

Coop options. The coop designs available now are overwhelming. I went high-end with my chicken coop, opting for the deluxe Carolina Coop model from Carolina Coops. This company uses animal-welfare practices in their design and materials choices, and the coops are far nicer than anything I could build myself.

While there are coops that I could see myself living in, all chickens really need inside a coop are protection from predators, protection from the elements, ventilation, nest boxes for them to lay eggs, and a roost for them to sleep on at night. Apart from the coop itself, chickens want room to roam—whether that's in an enclosure or free range—for their enrichment, foraging, and overall well-being. They also need a clean water source and a food source.

As long as you can accomplish these chicken needs, you can build your own coop, find one used (that's been cleaned and sterilized), purchase a kit, or bring in a preassembled coop and run.

The Carolina Coop, a 6 x 18-foot (1.8 x 5.5-m) masterpiece of chicken coop engineering.

Jacques' DIY coop is highly functional and affordable.

Coop bedding. Chicken waste is a rich resource for homesteaders. Managed well, it's not going to make a stink, which is important in all settings, and especially when the coop is near your house or someone else's. Choosing the right bedding is important here. You want bedding that's absorbent, made of natural materials that can be composted, and not slick when wet. I started out with a mix of pine shavings and industrial hemp, and now I'm just using industrial hemp. Part of the decision about the bedding choice you make will depend on what's readily available to you, as well.

Chickens won't leave droppings in the nest box, but they want bedding here, too. Long fibers, like straw, are ideal, because the chickens can make a nest and leave their eggs in a place that feels safe for them.

Following the deep-litter bedding method, like I do, means bedding only needs to be topped off every month or when you notice it's wet or smelly. Deep-litter bedding starts composting in place, and the good bacteria keep the "bad" bacteria in check. Remove the bedding and start fresh once a year or so. Very small coops will need the bedding to be changed more frequently.

Industrial hemp bedding plus a "deep litter" method means less hen house cleanouts.

EPIC TIP

MAKE ROOM

One of my reasons to keep laying hens is so I am assured my eggs are produced with animal welfare in mind. To me, this means no living conditions that you'd find in industrial egg production. A healthy and happy homestead flock needs space. These are Carolina Coops' space recommendations:

- Give each chicken 1 foot (30 cm) of space on the roosting bars.
- If your chickens will be going inside and outside the coop, you want 3 to 5 square feet (0.3 to 0.5 m²) of space per chicken.
- For coops that house chickens without outdoor access, have 10 square feet (0.9 m²) per bird.
- Plan for three to eight chickens per nest box.

project
BUILD AN OUTDOOR RUN

Materials

Wire cutters

Hand drill and auger or manual post-hole digger

6-foot (1.8 m) tall 4 x 4 wooden posts

Concrete mix

Cabinet screws

Gates, whether you build or buy them

Gate hinges (at least 2 per gate)

Gate latches (1 per gate)

Lumber of your choosing

6-gauge, powder-coated hog panels

Fence wire clamps with screws

The coop at the Epic Homestead came with a small, covered outdoor space for my six hens. I have all this room outdoors and know that the more space they have, the better.

This project outlines the materials and techniques we used to give the Epic Chickies triple their outdoor run space. The automatic coop door lets them in and out of their fully enclosed area during the day and closes itself at night, after the hens have gone inside.

The coop itself, the garden shed, the water-storage tank, and the property fence act as parts of this pen's enclosure. We built an attractive and durable 4-foot (1.2 m) tall fence to close off the rest. You can adapt this to any size of run you want at your homestead. Think of this as a jumping-off point.

The outdoor run addition tripled the space my hens have to roam.

Leveling the ground with a rake.

Building Instructions

1. Plan the fence. Measure the area you're fencing so you can get the right materials from the start. Consider how you access the coop and other things in the area of this chicken run, and plan the fence to keep these areas accessible. Measure your gate openings.

2. Prepare the area. Remove any wood chips, vegetation, or other obstructions that will get in the way of your fence posts. Level the area somewhat—this doesn't need to be perfectly level, but level enough so the bottom fence boards can be straight.

3. Drilling holes to set posts.

4. Setting posts in concrete and making sure they're plumb.

5. Attaching hog panel to the posts.

3. Dig post holes. Measure, mark, and then dig holes for the fence posts. A Power Planter auger attachment for a hand drill is very handy.

4. Set posts. Sink posts 2 feet (61 cm) deep so they stand 4 feet (1.2 m) tall. Level them, and pour cement around the base to make this fence really last. Learn from us, and let the concrete set before attaching anything to the posts.

5. Construct gates. While the concrete is setting, assemble the gates you need. You might just have one gate into the enclosure, and you might purchase that gate already assembled. I had two gates, and we built them using a simple wood frame, hog panels, and leftover fencing materials.

6. Install hog panels and lumber. This is a very choose-your-own-adventure project. I chose to purchase sturdy metal hog panels for this fence. I know the holes in these panels are large enough that a predator could squeeze through, but this fenced area is for daytime use only, and I don't have daytime ground predators at the Epic Homestead. I dressed up the finished look by using lumber leftover from the decorative wooden fence that we took down in front of the house: Three boards above the hog panel and three below.

Attach the hog panel using fence wire clamps with the provided screws, and attach the lumber using cabinet screws.

7. Hang gates. Level and hang the gate or gates where you need them. You may find, like I did, that the ground under the gate needs more leveling so the gate can swing freely.

8. Look for holes. Step back and think like a predator—or a chicken. Identify gaps or holes in the perimeter fence, and fix those. This will both keep out ground predators and keep in chickens. For my fencing installation, we saw a large gap under the garden shed, which serves as one side of the enclosure. Jacques simply secured a 2 x 6 board with some rebar as a patch.

9. Let the chickens test the design. Release the hens into their new space! Keep an eye on them for the first day or so, and they'll show you other areas that need more security.

10. Consider aerial protection. Hawks, eagles, ospreys, and other daytime aerial predators—left to their own devices—will come for your chickens. If you live in an area with aerial-predator pressure, you may need to block them. Bird netting or strips of reflective material strung from the fence to the coop can do it. Get creative to keep your chickens safe.

Adding planter boxes in front as chicken forage.

Strengthening the panel with fence wire clips.

The Epic Hens enjoying their newfound outdoor space.

The automatic coop door ensures I never forget to close up the coop at night.

This water bar features beak-activated watering nipples.

Coop Tech

There is no shortage of ways to use technology to make chicken keeping more efficient. I really appreciate the features I was able to add to my coop in line with high-tech, natural living. They've saved me time in managing the flock, money in reducing wasted feed, and worry in knowing they're protected.

The automatic coop door saves a massive amount of time for me and is good for the chickens. Unlike a lot of humans, chickens become active at dawn and are ready to eat, drink, and get on with their busy day of scratching and foraging. If you're planning on being outside every day when your chickens wake, I applaud you. If you're not, the automatic door will let them out while you look out the kitchen window with your coffee. Likewise, in the evening, the coop door needs to be closed at dusk to keep the chickens in and the predators out. The chickens habitually go inside and get on their roosts as the sun fades. This automatic door closes the coop when the sun goes down and saves you the hassle of having to be home at exactly the right time every day.

A rain barrel alongside the coop can catch the water from the coop roof and use it to water your hens. It's super convenient to have a water source right there, and you can even attach the rain barrel to an automatic waterer. (See Chapter 7, page 157, for more about rainwater catchment.)

No-freeze waterers are less of a need for me in San Diego, but my chicken keeping friends elsewhere rely on them. There are different models out there. The one that came with my Carolina Coop is set up with a rain barrel that acts as a water reservoir for the coop. A submersible heater keeps the rain barrel from freezing. Then a circulating pump sends water through a PVC pipe fitted with nipple drinkers.

Well-designed feeders help conserve feed, save money on wasted feed, and create less of the mess that would attract pests. I like the CoopWorx feeder because it holds two bags of feed and has eight feeding areas, so there's no fighting over who gets to eat. There are also feeders that release feed when a chicken steps on the platform. Chickens are smart and figure out this game quickly.

A multi-entrance feeder is a coop essential.

The egg hutch is easy access and one of the most fun parts of my morning routine.

Most of the time, scraps from the garden go to the hens.

Spent hen house litter makes perfect compost material.

Letting your hens roam the garden is free pest control.

Using the Outputs

The more you know about keeping chickens, the more sense they make on a homestead. Here are my favorite obvious and less obvious rewards I get from these birds:

- **Recycling food scraps.** These birds are fairly indiscriminate eaters. They'll clean up veggie scraps, spent garden plants, leftover yogurt, and more. Today's chicken breeds can't subsist on scraps alone, but the scraps make a nice supplement to their diet.

- **Compost.** When it comes time to clean out the coop—just once a year if you're using the deep-litter method—it's easy to scoop from the coop into the compost bin. Go back to Chapter 5, page 129, to read about how to safely compost chicken manure.

- **Eggs.** I've collected so many eggs from my six chickens that I've had to give them away, freeze them, dehydrate them, eat them—which is my favorite part—and otherwise preserve them. A tip to free up some fridge space as all of your greens and eggs come in at the same time: Unwashed, intact, previously unrefrigerated eggs can be kept on the countertop.

- **Pest control.** Running chickens through your garden while the garden is producing is a risky idea. Their waste is a food-safety hazard, and their scratching and pecking can damage plants. Once the garden season is finished, giving chickens the run of the place can reduce your insect pest population. In orchards, too, chickens work wonders. Don't get grossed out by this, but another homestead pest they love is mice!

CHICKENS VS. DUCKS

I've been Team Chickens for a long time, and I never thought I would be Team Ducks, but I'm coming around. They're both small, feathered birds, so you'd think they wouldn't be that different to care for, but they are.

Ducks differ from chickens in their shelter requirements. They don't roost, so their house doesn't need to be tall. But because they sleep on the ground, they're more susceptible to being nabbed by a predator reaching under their shelter. You have to fortify that area to keep out digging animals.

There's another big difference: *water*. You don't need a huge pond to keep ducks, but you do need at least a kiddie pool, and they will make a mess of it. It doesn't take long for the whole yard to become muddy. Chickens, on the other hand, only need water for drinking.

A lot of zoning ordinances don't address whether it's legal to keep ducks. They can be pretty noisy—you might call them enthusiastic—so they wouldn't be easy to hide from close neighbors.

Like chickens, ducks are great at controlling pests—especially slugs and snails. Their droppings are a food-safety hazard in the garden, but they cause less damage to plants because they don't scratch and peck like chickens.

The duck eggs are the biggest treat. Rich, large, and laid pretty regularly regardless of duck age and time of year, these are in a whole other category of egg.

Quail are small and quiet. They produce tiny, delicious eggs and are a source of rich, flavorful meat. A few friends of mine keep quail on the downlow, and I'm surprised quail haven't caught on as more of a go-to modern homestead animal.

These tiny gamebirds are relatives of the much-larger pheasants. There are more than 100 varieties of quail in the world. Coturnix, or Japanese, quail are larger and faster growing. By "larger," I mean they mature to about 10 ounces (283 g). They also lay about 200 eggs per year.

Baby quail are even smaller and more fragile than baby chicks. They can fit through very small spaces—tighten up that brooder enclosure—and can even drown in shallow water. Quail and egg-laying chickens have different nutritional needs, so they'll need their own feed, high in protein.

As quail mature, they begin to fly. They need a fully enclosed pen to keep them from flying out to find their wild cousins. They need just 1 square foot (0.09 m²) of ground space for every three birds in a coop, and they do well with long runs of about 6 square feet (0.6 m²) per bird for them to run and fly.

BEEKEEPING

Anyone without a garden might first call out honey as the reason to keep bees. Honey is a great reason, but it's not what I think of first. It's the way that bees make honey that really makes me want to keep bees.

About 30 percent of fruits and vegetables need insect pollination to produce. This includes your cucumbers, squash, melons, broccoli, apples, citrus fruits, and more. Keeping bees near your garden nearly guarantees they'll visit your plants.

While I see bees as an essential part of a homestead, I'm just starting to learn about and work with bees myself. Much of what I know about keeping bees comes from my friend Hilary Kearney, a San Diego beekeeper who you might know from Girl Next Door Honey.

There are a couple of things I know about beekeeping for sure.

First, you need a mentor. This is somewhat true for everything in homesteading, but it's especially true in working with bees. Bees are unlike anything else you'll work with. Your beekeeping mentor will save you the headache, and probably the sting, of having to rebuild your habits and your hives. A mentor will help you learn to listen to your bees, interpret all that you're learning, and find the beekeeping methods that work for you.

Second, beehives are among the last of the projects to install on a homestead, especially in a space as small as the Epic Homestead. One of the greatest misconceptions about bees is that they're mean. They're not mean, they're just protective. Bring in hives when you're sure you know where to site them, so you don't have to move boxes and disrupt the bees. The bees will be less aggressive and more productive when they can get themselves established and come and go with little disruption.

My first time learning to harvest honeycomb.

The Langstroth hive is, by far, the most common.

A bit fancier, the Flow Hive helps automate the process.

The top bar hive is the oldest and most natural method.

Choose Your Hive

There are a dozen different types of beehives in use the world over. I'm most familiar with three common types in the United States: Langstroth, Flow, and top bar hives. I encourage you to also learn about others, including Warre, Apimaye, and Layens.

Langstroth hives are the classic stacked-box setups. In the mid-1800s, the inventor, Lorenzo Langstroth, figured out "bee space" dimensions, which is the exact amount of space the bees need to move throughout the hive. He really figured out a lot about how bees operate to create his design.

A Langstroth is the standard starter beehive, but that doesn't make it simple. There are lots of options for how to use a Langstroth hive, from the size of the boxes (eight frames versus ten frames and medium supers versus deep supers) to the type of frame (foundationless frames or frames with comb already in place).

Flow Hives have been in use for about a decade, and because I'm all about innovations that bring modern technology into homesteading, they need to be mentioned here. This is a specially designed box to make honey extraction easier. You can start with a whole Flow Hive setup or you can add a Flow Hive box to a Langstroth setup.

The Flow Hive box includes frames with clear ends and a side window so you can see when honey is ready to harvest. When the frame is full, instead of removing it, disrupting the bees, and running the comb through an extractor, you turn a key to uncap the honey cells, and the honey flows out of a spigot on the side. Other than this difference, the management is identical to that of the Langstroth hive.

Top bar hives are the oldest hive design and most closely mimic a hive in nature. These may be the most common hives used outside of North America. Instead of using a four-sized frame, the bees build their comb from a single horizontal bar. Bees move between hanging frames to build their comb, set their brood, and make their honey.

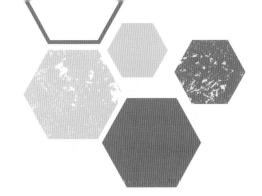

HIVE TYPES PROS AND CONS

LANGSTROTH HIVE

PROS
- With standard measurements, you can order Langstroth parts from any bee store.
- If you need help, you're likely to find a beekeeper with Langstroth experience nearby.
- You can add and take away boxes in different sizes as needed.

CONS
- These boxes are heavy—60 to 100 pounds (27 to 45 kg) when full, depending on the size box you're using.

FLOW HIVE

PROS
- This is the easiest, least messy honey extraction method there is.
- You don't have to disturb your bees for the honey harvest.
- It integrates with Langstroth hive equipment, which is easy to find.

CONS
- The comb foundation is made of plastic.
- It's more expensive, as technology tends to be.
- There's more of a tendency for beekeepers to overharvest the honey and leave the bees without enough reserves to get through the lean times.

TOP BAR HIVE

PROS
- Top bar hive management is easier on your back. You deal with the frames one at a time instead of in heavy supers.
- You can easily build your own top bar hives to fit your size and space needs.
- You can harvest a lot of beeswax from these hives.

CONS
- The combs are less stable and will more easily break, losing valuable beeswax and honey.
- The bees have to use their energy to rebuild the comb after each extraction and may produce less honey as a result.
- There's a greater learning curve to working with top bar hives than with Langstroths.

Planning Your Hives

Bees are foragers and can fly 2 miles (3.2 km) or more in search of nectar and pollen. You can't fence them in like you do chickens. During nice weather, bees come and go all day long, and with 10,000 bees in a hive, the flight path is a busy place. Direct their hive opening so they fly into your yard instead of directly into a neighbor's yard. You may have to redirect their flight path by planting a hedge or putting up a tall fence a few feet in front of the hive. They'll be forced to fly higher, away from people who might panic at the sight of a bee.

Flowering plants are bees' primary food source. In some urban areas, it's hard for bees to find enough to eat. Plant as many varieties of plants that flower throughout the season as possible.

A dedicated beekeeper inspects a row of hives.

One hive of bees will drink as much as 1 quart (1.1 L) of water a day in the summer, and you need to provide that water for them. A shallow birdbath with consistently fresh water will do. Unfortunately, if your neighbors also have a swimming pool, it's hard to stop them from visiting that pool instead. Be sure to have your water source in place when your bees arrive so they can create a habit of visiting the water intended for them.

They want to be in a quiet area of your lot. A corner away from a busy road that doesn't get a lot of activity is ideal. You can also, with a flat roof like mine, place hives on a rooftop. Just be sure the roof is rated to hold the weight.

I hope you'll consider adding mini livestock to your own epic homestead when the time is right. Next, let's focus on another way to improve your self-sufficiency—through proper food preservation and storage.

PERMITS AND ZONING

You read about the challenges of working with zoning regulations in Chapter 1. For mini livestock on a homestead, you really want to get this right, because you're dealing with animals and insects, not just plants and property features. The stakes are higher.

Chicken keeping is highly regulated. The number of chickens, whether you can have roosters, the size and location of a coop, required permits, and more are outlined in many city and county ordinances.

You'll usually find similar regulations for bees as you do for chickens. There's plenty of misinformation circulating out there about chickens, but with bees, it's even more of a challenge to educate neighbors—and zoning commissions—about their harmless nature and community benefits.

What's often not outlined in the ordinances are other types of poultry and game birds. Ducks and quail are rarely mentioned. This doesn't mean that you can get away with keeping them carefree—but it doesn't mean you can't get away with keeping them, either.

There are two schools of thought: The first is to go ahead and get the animals and see what happens. You can claim that you aren't doing anything wrong because they're not regulated. Sometimes this works. The second is to bring your interest in keeping these birds (or bees) to the zoning board's attention. You can then work with them to create fair ordinances. This approach works better when you have the backing of an organization or other homesteaders who also want to keep mini livestock.

Quail may carry with them an extra layer of permitting. If quail are native wild birds in your state, the wildlife conservation office probably wants to keep tabs on them. This is to ensure your bevy of quail doesn't pass diseases on to the wild population, and vice versa.

There are even more regulations if you're planning to make money from any of your dealings with mini livestock. Business permits and zoning are completely different, so be sure to research those if you're taking that route.

PLANT WITH MINI LIVESTOCK IN MIND

A diversity of food sources enriches the diet and health of bees, poultry, and gamebirds. Here's a very short list of perennial plants that bloom throughout the seasons that you might put on your homestead for your winged friends. Choose a few that are native to your area.

- Bee balm
- Black-eyed Susan
- Catmint
- Chives
- Columbine
- Echinacea
- Goldenrod
- Hardy geranium

- Japanese anemone
- Lady's mantle
- Milkweeds
- Peony (single types)
- Russian sage
- Shasta daisy
- Spearmint
- Yarrow

Learning the forgotten skills of food preservation opens up an entirely new world of flavor, texture, and nutrition to life at your homestead.

chapter nine

FOOD PRESERVATION AND STORAGE

IN MY CHILDHOOD, I ate what I consider to be standard 1990s food in America. My mom cooked, but there wasn't a lot of knowledge of where the food came from or the nutrient density or quality of it. I'm not putting any blame on my mom here. I think the mass food-corporation marketing of the 1990s and an out-of-whack food pyramid got the best of a lot of families in the United States. Today, of course, where and how food is grown is something many of us talk about all the time.

What gets me excited about all the food I'm producing at the Epic Homestead—the fresh eggs, the fruit trees, all these vegetables and herbs—is knowing that the quality of this food is so much better because of how it's grown. The eggs are better because of how the hens live and eat. The vegetables are grown in soil I'm improving so they're more nutrient dense.

The freshness factor is so crazy when you're growing stuff on your own homestead—even when that homestead is a 400-square-foot (37 m²) apartment. Everything is fresher than anything I can find in a store. And then I get to preserve and capture unique flavors that I didn't even know existed, like new-to-me favorites, loquat jam and loquat crumble.

For me, storing this food properly until I can eat or preserve it is important. I'm bummed when I accidentally let it go and have to give it to the chickens, worm bin, or compost—but I do give myself some grace because this isn't totally a waste. Preserving this food means I get to enjoy it no matter the time of year. What's in season from month-to-month changes quickly, and then that produce is gone, unless you've managed to freeze, can, dehydrate, or freeze dry it. I want to actually savor these flavors after all the work that I've put into creating them.

 Scan for video

A typical harvest from the early spring garden and orchard.

Leafy greens need high humidity to store well in the fridge.

STORAGE

When you think about it, food starts to decline from the moment you harvest it. Leaves of kale, melons, and sprigs of rosemary were alive and thriving on the plant, then you remove it from its source of life, and it becomes a race against time to consume it. In the peak of garden season, you might struggle to keep up with your harvest, finding you've let a few peppers get soft on the kitchen counter or a bag of green onions wilt on a refrigerator shelf.

Storing your vegetables properly helps keep their quality so you can enjoy them for a longer time. The right kind of storage is also important if you plan to freeze, can, or dehydrate your harvest. You want to start with the best-quality produce possible to maintain the taste, texture, and color in preservation.

Storing Leafy Greens

In a plastic container in a crisper drawer, greens will stay green for a week or longer. Vegetables in this category include kale, spinach, Swiss chard, collard greens, and more. Arugula and tender lettuces need to be used soonest.

Storing Root Vegetables

In general, remove the tops. Store the tops as you would leafy greens. Store the roots in a plastic bag in the crisper drawer. Kept at high humidity, roots will last for weeks. Some root vegetables get sweeter with storage.

Onions. If cured by letting the harvested bulbs sit in a warm, dry, well-ventilated spot for three weeks, they can keep on a countertop or pantry in a cool room. If fresh, bag them and put them in the fridge. They could change the taste of other things in the fridge if they're especially strong. Double bag these onions, or put them inside a sealable container.

Potatoes. If cured, leave mature potatoes on the counter in a dark place so they don't develop green spots, which are toxic. To cure potatoes, keep them at 45°F to 60°F (7.2°C to 15.6°C) and high humidity for two weeks. If they are new potatoes (those harvested from still-green plants), keep them in a paper bag in a crisper drawer.

Sweet potatoes. Start by curing sweet potatoes by keeping them at 85°F (29°C) with high humidity for a week. Then store them at room temperature for months. Sweet potatoes will develop black spots if refrigerated.

Garlic. Cure garlic at temperatures around 80°F (27°C) for two weeks. Braid them, or spread them out on a wide, flat surface. Be sure they have good airflow and no moisture. Keep garlic bulbs in a pantry for up to a few months. Put them in the fridge for longer keeping.

I tie garlic into bunches to cure, then store, for the long-term.

EPIC TIP

CURE YOUR STORAGE VEGETABLES

Curing maximizes the storage life and quality of some vegetables. Onions, garlic, potatoes, sweet potatoes, and winter squash all benefit from curing. By exposing these vegetables to a certain level of heat and humidity for a period of time, you're concentrating their sugars and thickening their skins. You'll find that even minor blemishes tend to heal over. Properly cured vegetables last longer in storage than their uncured counterparts.

You might need to be creative to find a place to cure vegetables. Try a greenhouse or a small room with a space heater and humidifier, depending on the conditions you're after.

Storing Tomatoes

Tomatoes lose flavor when refrigerated. It's best to leave them on a countertop. If they'll go bad before you get to process them, core and put them in freezer bags. You can process them from frozen when you're ready.

Storing Winter Squash

Cure your winter squash and pumpkins, at 80°F to 85°F (27°C to 29°C) and high humidity. Keep them in a pantry or on a countertop after that.

Storing Herbs

To keep herbs at their freshest, store them with their cut ends in a jar of water, loosely covered with a plastic bag. Change the water every day. Most herbs can stay in the fridge like this. Basil is very cold sensitive and may need to sit on the counter.

Storing Fruits

Keep fruits and vegetables in separate drawers in the refrigerator. Fruit releases ethylene, a gas that makes produce ripen, which can cause vegetables to go bad sooner.

Apples and peaches. Store at room temperature for up to a week or in the fridge or a cool, dark place for longer. They'll take on flavors, so don't store them near onions.

Apricots, berries, cherries, figs, and grapes. Keep these in the fridge in a perforated plastic container.

Bananas, citrus fruits, pomegranates, and persimmons. Always keep these at room temperature.

Melons. Leave a melon at room temperature until you cut it open, then put it in the fridge.

Pears and plums. Put these on the counter to ripen, then keep them in the fridge.

Storing Other Vegetables

Asparagus and celery. Store in the fridge with their cut ends in a jar of water.

Brussels sprouts, green beans, peas, and okra. Put them in a breathable bag, like a paper bag or a perforated plastic bag, in the crisper drawer.

Cabbage. Don't even bother with a bag for storage. Just keep it in the crisper drawer, and peel off the outside layer of leaves before you use it.

Cucumbers, peppers, zucchini, and eggplant. These don't love very cold temperatures. Keep them in your fridge door in a perforated plastic bag. If you have a 55°F (13°C) space elsewhere in the house, they would do well there.

MECHANICAL DEHYDRATION

The definition of dehydration is the removal of water, and that's what's happening when you dehydrate food for preservation. Food is prone to mold, yeast, and bacteria growth because of its moisture content, so if you remove the moisture, you remove the likelihood of spoilage.

A little later in this chapter, you'll read about air drying foods. Using a dehydrator for mechanical dehydration is the more high-tech, modern way of food preservation. It's certainly faster, and it can result in better-quality dehydrated foods. A mechanical dehydrator uses low-humidity air, circulation, and a consistent temperature to dry out foods.

You may need to follow different steps, depending on what foods you're dehydrating, but there are a few principles that apply to all foods:

- The more surface area exposed to the hot dehydrator air, the faster the food will dehydrate.

  Spreading out the material as evenly as possible across the dehydrator tray will speed the drying time.

- Before packaging anything for storage, test a few pieces for the telltale snap of a well-dried vegetable and the crumbling of a dehydrated leaf or flower. Fruits are considered dried when you can't squeeze moisture from them and they're tough and pliable. The last thing you want is for some of the material to be dry and some of it to still have moisture in it. That little bit of not-dry material can cause the whole container to mold. When in doubt, give the food another half-hour on the dehydrator, let it cool, and test it again.

Turning homegrown herbs into flavorful spices is one of my favorite ways to preserve the harvest.

Break your fruits and veggies down into uniform pieces for a smooth dehydration experience.

Using a spice grinder will get you finer herb grinds.

The final product.

Dehydrating Herbs

Preparing the herbs to go into the dehydrator varies according to the type of herb. For larger-leaved herbs, like Genovese basil and common sage, remove the leaves from the stem, and lay the leaves flat on the dehydrator tray. For herbs with small leaves, like rosemary or lime basil, lay whole sprigs of the herb on the tray. Roots and rhizomes, including ginger and burdock, need to be uniformly sliced or grated for any hope of dehydrating them.

Dehydrate leafy and flowery herbs on a very low temperature. I dry my calendula and chamomile at 95°F (35°C) for three to four hours. Delicate leafy herbs, like basil, may take less time. Herbs with woodier stems and sturdier leaves, like rosemary and oregano, will probably take longer. You might have to raise the temperature for roots and barks—herbs that have a tougher, thicker body.

Use these herbs in your cooking and in teas. I like to break the dried herbs into smaller pieces using a spice grinder. You can use a small food processor or mortar and pestle, though the herbs won't come out as finely ground.

Dehydrating Vegetables and Fruits

I dehydrate my loquats every season, as they come in hot and heavy for about a month and then are gone until next year. Fruits won't need to be blanched, but they'll need to have their pits removed.

You can dehydrate vegetables from their natural state, but blanching them generally keeps their flavors and colors more vibrant. They may dehydrate more quickly, too.

Some dehydrated vegetables and fruits are ready to eat as snacks—think about dehydrated apple slices and cabbage chips. Others need to be rehydrated or added to brothy foods for cooking, like cubed carrots, potatoes, and green beans.

Dehydrating Mushrooms

A run through the dehydrator is the best way to preserve mushrooms. I have a ton of foraged morels in the pantry as I write this. Dehydrating maintains their flavor, and they'll keep for a long time. Once dehydrated, you can grind mushrooms into a powder for a rich umami seasoning. You can also leave them intact and rehydrate them for any recipe.

Dehydrated loquats make the perfect snack for a long hike.

Perfectly seasoned, crispy cabbage chips.

recipe
CABBAGE CHIPS

You've heard of kale chips, and cabbage chips are just as good. Use your favorite seasonings here.

INGREDIENTS
1 head cabbage
Salt
Garlic powder
Smoked paprika

STEPS

1. Blanch cabbage leaves for 90 seconds in boiling water and then dunk for 90 seconds in an ice water bath. Drain the cabbage, and slice the leaves into more manageable pieces.

2. Place cabbage slices on dehydrator trays in a single layer. Sprinkle with seasonings.

3. Set the dehydrator to 140°F (60°C) for 10 to 12 hours, until crispy.

4. Store in sealable bags or containers in the refrigerator or freezer for up to a year.

Make sure the leaves don't touch each other.

EPIC TIP

DEHYDRATE IN THE OVEN

I didn't have a dehydrator when I started growing my own food, but my home had a makeshift dehydrator already—an oven. Ovens are less efficient than mechanical dehydrators, but in a pinch, they'll do. Convection ovens are ideal because the air circulation helps herbs dry more evenly.

Thick stems and roots can dehydrate at 170°F (77°C) for 7 to 10 hours. Leafy herbs need lower temperatures: 110°F to 130°F (43°C to 54°C) for 3 to 4 hours.

Leave the oven door cracked just a bit to let out the moisture that's escaping from the herbs. You want to dehydrate, not roast.

Dehydrating Eggs

Dehydrating eggs is more involved than freezing them (see page 206). Start with scrambling the eggs. Next, you're going to dry cook them—no butter or oil, just eggs in a pan over extremely low heat. Keep the eggs from sticking to the pan by slowly working them with a spatula as they thicken up. When the eggs are fully cooked, transfer them to a dehydrator sheet, and spread them out as thinly as possible.

I dehydrate my eggs at 140°F (60°C) for 18 hours. Your time and temp may be slightly different, depending on the dehydrator you're using. When fully dehydrated, eggs won't be rubbery at all but will crack when bent.

From here, I powder the eggs in a spice grinder. You could use a mini food processor or coffee grinder. Store the dehydrated, powdered eggs in an airtight jar with a silica packet or oxygen absorber. Leave the jar in a cool, dark place for up to two years.

The American Egg Board points out that eggs are 75 percent water. This means to reconstitute eggs from a dehydrated powder, you need to add three times the water back to the powder by weight. Let's say you are using ½ ounce (14 g) of powder. Multiply this by 3 to find you need 1½ ounces (43 ml) of water. Blend this together before adding to your recipe.

I recommend reconstituting some eggs, cooking them like a scramble with oil or butter, and adding salt, pepper, and other seasonings. Spread this on some toast with homegrown pickled red onions.

Cook eggs in a pan with zero oil or butter.

After dehydrating, grind in a spice grinder.

Store in a Mason jar.

Laying out sliced ginger for homemade ginger powder.

EPIC TIP

PERFECT YOUR DEHYDRATOR ROUTINE

Every dehydrator is different, so it'll take a little time for you to get to know yours. Here are a few tips to get you started.

- Read the manual for suggested drying times. This is a starting point. Your house may have different humidity and temperature conditions than whatever lab the dehydrator was tested in.
- Check message boards for others' experience. Search for your dehydrator model and the foods you're drying to see if anyone has gone through this trial-and-error period before you.

- Take notes on your experience. For each run, write down what you dehydrated, how you prepped it, how long it took, and the temperature you used. This log will help you get the best results over time and serve as a reminder for how to dehydrate specific foods next season.
- Separate the items you're dehydrating by tray. Partly, this will help you remember which is which. Also, when the basil dries faster than the calendula, you can remove the basil tray without disturbing the calendula.

USING DRIED HERBS

Dried herbs are more potent than fresh herbs. For every 2 teaspoons (8.4 g) of fresh herbs, you would substitute ¾ teaspoon (3 g) of dried herbs or ¼ teaspoon (1 g) of powdered herbs.

AIR DRYING

Without a mechanical dehydrator, all is not lost. Many herbs—and a few vegetables—dehydrate well in a dry, ventilated space. Attics, spare rooms, screened-in porches, and spaces near cooking fires were used for air drying long before mechanical dehydrators came about.

Your success with air-drying foods depends largely on the humidity in the space. With a fully air-conditioned indoor space, you can quickly remove moisture from herbs and some vegetables.

You'd think that drying herbs and vegetables in the sun would speed the process. It might, but sun-drying bleaches the color and weakens the flavor, so I don't recommend it.

Air Dry Herbs

Tie herbs into bundles or place them in one layer in a paper bag to hang, or spread them in one layer on a screened tray. In one to two weeks, you can have dried herbs. Lemongrass, tulsi, sage, rosemary, thyme, and many more aromatic herbs can be air dried. Basil does better in a mechanical dehydrator, as do some of the more mucilaginous herbs, like borage.

Hang-drying herbs is the easiest way to get into preservation.

Air Dry Vegetables

Being thicker and with more water content than herbs, many vegetables will end up rotting before they air dry. One exception is peppers, and specifically hot peppers. If you spend any time traveling outside the United States, you know that strings of chili peppers are a common sight. In the Southwest, as well, these strings—ristras—are everywhere. There's an art to creating traditional ristras. You can learn that technique, or create your own simple ristra style with a needle, thick thread, and peppers. (Be sure to wear rubber gloves!) After the peppers are dry, grind them into red pepper flakes or make a hot sauce.

Air-drying Shishido peppers for storage.

FREEZE DRYING

Now we've come to high-tech food preservation. When I first learned about freeze dryers, I thought I probably didn't have a use for them. The freeze-dried astronaut ice cream on the space-museum field trip was great, but freeze-dried ice cream wasn't something I thought I needed in terms of sustainability. Freeze dryers cost a couple thousand dollars, and for what I was growing in my small space, I couldn't justify the cost.

Moving to the Epic Homestead—producing all this food—a freeze dryer made more sense. It also helped that I had a taste of a friend's freeze-dried strawberries and realized this was about much more than ice cream.

The Freeze-Drying Process

When you experience the freeze-dryer technology, you begin to understand the expense behind it. The door has a rubber gasket that creates a tight seal when locked. This is important for the pump to create a vacuum. The machine first freezes the food. Then it heats the food, and because of the pressure of the vacuum, the moisture in the food evaporates. This process of going straight from a solid to a vapor is called sublimation. Freeze-dried foods last longer and are more lightweight than dehydrated foods because 98 percent of the moisture is removed this way.

EPIC TIP

MAKE YOUR OWN SPICE BLENDS

By the end of the season, you can have jars upon jars of various homegrown herbs. Used separately in the kitchen, these make for great additions to your meals. Combined into spice blends, the enjoyment of these herbs is magnified. A favorite spice blend of mine is an Italian seasoning with sage, oregano, basil, and rosemary from the garden.

Freeze Dry Fruits and Vegetables

Everything coming out of the freeze dryer tastes just like it did when it went in. For naturally juicy fruits and vegetables, it's an all-around pleasant eating experience. Tomatoes, carrots, peppers, and strawberries taste garden fresh!

Freeze-dried fruits are great snacks. Some vegetables are, as well, but in my experience, freeze-dried vegetables are better when reconstituted in soups and casseroles.

Prep fruits and vegetables just as you would for dehydrating. When running different types of fruits and vegetables in the freeze dryer at the same time, match them with their expected drying time. Unlike a dehydrator, you can't just start and stop the machine to remove trays. They're in the freeze dryer until the cycle is finished.

Freeze Dry Whole Meals

For camping, preparing for time off grid, or a way to have some meals on hand that don't require much storage space, you can freeze dry soups and casseroles with great results. Rehydrate soups and casseroles by mixing in warm water at ⅓-cup (78 ml) increments to reach the desired consistency.

Storing Freeze-Dried Food

Any time spent in the open air gives the freeze-dried foods a chance to reabsorb moisture. As you take the finished product out of the freeze dryer, put it right into a sealed container. Add a food-safe silica packet for good measure. Freeze-dried foods are shelf stable and will keep for years in your pantry.

Setting up the futuristic-looking freeze dryer.

Taste-testing freeze-dried tomatoes.

Strawberries are hands-down the best fruit to freeze dry.

The homestead pantry is filled with Mason jars like these.

CANNING

Canning is one area of food preservation that I'm determined to learn more about. Preserving food in reusable jars that are shelf stable is a major move toward sustainability. The process requires careful attention—you really have to follow the canning guidelines to do this safely. There's a larger time investment in canning than in the other preservation methods, but once you get rolling, you can preserve a lot of food in one day.

Canning works in a few steps. First, the water bath heats foods to the point that bacteria, molds, and yeasts are destroyed. It also stops enzymes from changing the color, texture, and taste of foods. At the same time, the air is drawn out of the jar. When the jars are out of the water bath and cooling, you hear that satisfying "pop" sound, which means the lid has sealed the jar so air and microorganisms can't get inside. All these steps together preserve your canned foods.

Produce, meats, syrups, jams, and soups are all options for the canner, as long as you follow approved recipes and canning methods. Start with fresh, good-quality foods to begin with. You can use less-than-perfect produce, but cut out the soft spots, mold, and other blemishes before processing.

Canning Tech

Since the late 1700s, homesteaders have used canning to preserve foods. The canner was invented in the mid-1800s, and honestly, not a lot has changed in canning technology since then. Techniques have been perfected, and the science behind them has been proven, but we're still using big pots of water and sealed glass jars to can food.

Here's a basic list of the canning tools to get started:
- Water-bath canner
- Steam-pressure canner
- Glass jars meant to withstand the high heat and pressure of canning
- New metal jar lids or reusable plastic jar lids with rubber seals
- Jar bands
- Jar lifter
- Wide-mouth funnel
- Bubble remover
- Multiple timers to keep track of everything cooking and processing

Canning Safely

Food safety and kitchen safety are so important here. There's no way to guarantee the food you've canned is free from bacteria, even if it smells and looks normal. The best you can do is follow the advice of experts. This includes what you'll find in the *Ball Blue Book Guide to Preserving* and information coming from university cooperative extension services.

This is by no means all that you need to know, but here are a few tips for safe canning:

Always follow approved canning recipes. Fooling with pH levels is different than tweaking an adobo recipe. Even if you're not the type of person to follow a recipe in the kitchen, you want to get this right by following the recipes that have been tested by the experts.

Use new lids every time you can.

LOW-ACID FOODS, HIGH-ACID FOODS, AND ACIDIC INGREDIENTS

LOW-ACID FOODS	HIGH-ACID AND ACIDIFIED FOODS	ACIDIC INGREDIENTS
Asparagus	Apples	Bottled lemon juice
Beets	Apricots	Citric acid
Carrots	Blackberries	Vinegar (5% acidity)
Corn	Lemons	
Green beans	Peaches	
Lima beans	Pears	
Meats	Pickles	
Okra	Plums	
Peas	Raspberries	
Prepared foods	Sauerkraut	
Spinach	Strawberries	
Turnips	Tomatoes	

Know the difference between low-acid foods and high-acid or acidified foods. Low-acid foods have to be canned in a steam-pressure canner, at a higher temperature, to kill all of the bacteria and toxins that are naturally in the food. High-acid foods have naturally high acid levels, and acidified foods have some kind of acidity added to them before canning. These can process in a water-bath canner, at a lower temperature. When you make foods acidified—adding acidic ingredients to foods to bring down their pH—you can treat these as high-acid foods. This is the case with pickles. Cucumbers on their own are not high-acid, but put them in an acidic vinegar brine, and they're a high-acid food. See the Low-Acid Foods, High-Acid Foods, and Acidic Ingredients chart at right for more examples.

Use new metal lids. Jars are reusable, and the metal rings you use on the jars are reusable, but metal lids are single use. They might not seal properly if you try to use them a second time around. Check out reusable plastic lids with rubber seals, if you're looking for a lid with a longer lifespan.

Adjust your canning time for your altitude. Because air pressure is lower at higher altitudes, this affects the temperature where water boils, and you need to process jars for a longer time and at higher pounds of pressure.

Inspect your jars before you get started. Don't use jars that are chipped or cracked, because they can explode when they hit the hot water.

FREEZING

Freezing is an easier means of preservation than canning, and you get a similar result in the end. The limiting factor for freezing is having enough freezer space for all that you want to save. I added a chest freezer to the Epic Homestead to make room for my harvest and so I can source meat in bulk from a small local farm.

Freezing Vegetables

Freezing vegetables isn't always as straightforward as you might think. Depending on the vegetable, you might have to do some prep work before putting it in the freezer.

Freezing vegetables usually begins with washing and chopping or shredding. Vegetables that you intend to use as a purée—pumpkin is the first one I think of—should go ahead and be cooked and puréed as normal.

Blanching comes next. This is a quick dip in boiling water or time in a steam bath, followed by cooling in very cold water. Blanching stops the enzymatic action that makes vegetables lose flavor, color, and texture. Different vegetables require different blanching times—no more than a few minutes. The *Ball Blue Book Guide to Preserving* is a great source for blanching times—and for all kinds of other preservation advice.

Drain the vegetables, and then pack them into freezer-safe bags and containers.

Freezing Eggs

Vegetables are not the only homestead product that you can freeze. There's meat, of course, but have you thought about freezing eggs? When the Epic Chickies are giving me five to six eggs a day, I get overwhelmed by them all. I don't want to waste a single egg, but I just can't eat thirty-five eggs a week. Then there are the times of the year when their laying slows to nearly nothing, and I wish I had some eggs. The happy medium is freezing the eggs.

In this tried-and-true egg-preservation method, you blend the eggs just to combine them, pour them into silicone muffin cups, and freeze them. When they're solid, pop them out of the cups into freezer bags, and you'll have eggs at the ready. Frozen eggs will keep for six months.

Store raw eggs in the freezer.

A frozen egg puck is perfect for breakfast.

You can get more specific here and freeze eggs in certain quantities. For example, blend up just two eggs at a time and pour that amount into one muffin cup so you know, when it's time to thaw and use it, that you're using two eggs. I tend to just blend a bunch together, fill the muffin cup, and eyeball it later.

Thaw frozen eggs in the refrigerator or in a sealed plastic bag under running cold water, and use them only in dishes you're thoroughly cooking.

OTHER PRESERVATION METHODS

Pickling, fermenting, and candying or making syrups are also pretty easy ways to hold on to your produce for a while longer. The best part about all of these is the flavor combinations you can come up with. Just add an herb or spice, or swap out one vegetable for another. You can be much looser with recipes here than you can when canning.

Quick Pickles

Quick pickles, or refrigerator pickles, are my go-to preservation method for vegetables. It's just so fast and easy. It's hard to mess them up!

Carrots, radishes, turnips, cucumbers, kohlrabi, and cauliflower make excellent quick pickles. Add to them whatever herbs and seasonings you have on hand, including some garlic.

Whipping up a quick brine.

recipe
BASIC QUICK-PICKLE BRINE

No matter what you're quick pickling, the basic brine recipe is the same.

INGREDIENTS
3 cups (710 ml) white vinegar or apple cider vinegar
3 cups (710 ml) water
3 tablespoons (38 g) non-iodized salt
2 tablespoons (26 g) sugar

HOW TO
Heat this until the salt and sugar dissolve, and pour over vegetables and herbs you've packed into a clean mason jar. Let it cool, put a lid on it, and keep it in the fridge. The longer the vegetables sit, the more pickled and melded the flavor. Eat these in a few days or a few weeks.

The amount of brine you need depends on the amount of vegetables you're pickling. You want your veggies to be covered by liquid in the jar. Keep extra brine in the refrigerator for your next pickling recipe.

The flavor of homegrown pickles can't be beat.

This cabbage is perfect for kraut, as it didn't grow as tightly as I wanted it to.

Fermentation

Fermenting foods is a giant science experiment that produces foods that happen to be full of probiotics, which are live cultures that aid in digestion. Lacto-fermentation converts the sugar in your ingredients into acids, gasses, and alcohol. *Lactobacillus* bacteria lowers the pH, making the food more acidic and less hospitable to other types of bacteria. It's a fascinating process to watch.

To get started, fermentation requires non-iodized salt, a vessel that allows limited exposure to air, and a way to weigh the vegetables so they stay submerged in the liquid.

While a sealed jar limits exposure to air, these don't always work for fermentation. As gasses are produced in the fermentation process, they need to be able to escape. Yes, you can use a sealed jar that you will burp every day, but the one day you forget to burp it, you could run into a burst jar, wasted produce, and a giant mess. An airlock lid is a smarter idea, because it allows gas to escape but prevents air from entering. To keep your ingredients submerged, you can purchase special fermentation weights or get creative and make your own.

I've used the Kraut Source fermentation kit from the time I started fermenting my own foods. I didn't have to figure out what jar, weights, or airlock I needed in the beginning—the kit made it easy. You can assemble your own kit, too, of course. You do want to keep your fermentation tools in the same place so they're easy to put your hands on come harvest time.

recipe
FERMENTED KRAUT

INGREDIENTS

1 2-pound (1 kg) head cabbage

2 tablespoons (30 g) plus 1 teaspoon (5 g) non-iodized salt, divided

4 carrots

1 tablespoon (15 g) caraway seeds

6 juniper berries

1 quart (1.1 L) sized mason jar

STEPS

1. Trim away loose outer cabbage leaves. Cut cabbage head in half, and remove the core. Thinly slice the cabbage.

2. Put cabbage in a large, stainless-steel bowl. Sprinkle with 1 tablespoon (15 g) salt. Massage salt into cabbage for 5 minutes to release moisture.

3. Shred carrots, and add to the bowl. Add 1 tablespoon (15 g) salt and spices, and mix together.

4. Transfer vegetable mixture to jars, and pack it down. Leave 2 inches (5 cm) of headroom.

5. Pour in any liquid that was released from the cabbage. Cover the jar with a weight and a breathable lid, and let the mixture sit at room temperature for 24 hours to release additional liquid. If liquid doesn't cover the vegetables, make a brine of 1 teaspoon (5 g) salt to 1 cup (237 ml) of water. Let the salt dissolve, and pour into jar to cover.

6. Store your kraut in a dark, room-temperature space for a week to 10 days. Taste it, and see if you like it or if you want the flavors to mellow more. The kraut might taste salty if it still has fermenting to do. When it reaches the right flavor, put a solid lid on the jar and transfer it to the fridge.

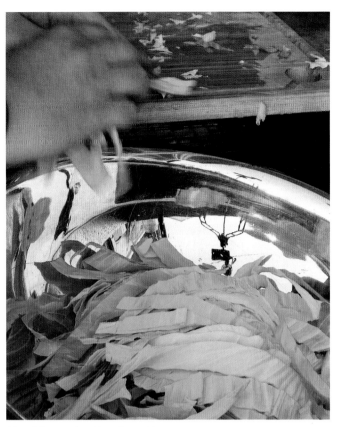

Shredding and salting the cabbage.

Keep the finished kraut below the surface of the liquid.

Syrups and Candied Fruits and Vegetables

On the sweeter side, you can make candied fruits and vegetables, which are actually two special preserved items in one. You end up with both the candied food and the syrup that was a result of making the food. Candied ginger is next level, but my personal favorite is candied jalapeños.

Sugar is a natural preservative, and candied foods require a lot of that. The ratio is typically twice as much sugar as liquid. The idea is the syrup will permeate the food matter, and the food will impart its flavor into the syrup. Syrups and candied foods need refrigeration or proper canning—these aren't shelf stable.

Dicing up a batch of jalapeños.

Pouring in the sweet, spicy slurry.

COWBOY CANDY (CANDIED JALAPEÑOS)

INGREDIENTS
1½ pounds (0.7 kg) jalapeños
3 cups (710 g) sugar
1½ cups (355 ml) apple cider vinegar
½ teaspoon turmeric powder
½ teaspoon chili powder
½ teaspoon onion powder

STEPS
1. Wear gloves to handle peppers. Slice jalapeños into ¼-inch (6 mm) rounds. Remove seeds if you want less-spicy cowboy candy.

2. Mix all ingredients except jalapeños in a pot, and bring to boil. Turn down heat and let simmer to thicken, about 6 minutes.

3. Add jalapeños, and simmer for about 7 minutes. Strain and reserve syrup.

4. Pack jalapeños into sterilized mason jars, leaving ¼-inch (6 mm) headroom. Pour syrup over to cover.

5. Allow these to sit in the fridge for a few days to a few weeks before eating. The longer they sit, the more the flavors will meld into a spicy-sweet treat. Keep in fridge for 3 to 4 months.

6. Put the candied jalapeños on crackers with cream cheese for epic appetizers. Use the syrup for cocktails and marinades.

CONCLUSION:
A New Chapter Awaits You

The joys of a homesteader's life make all the harder parts worth it.

The world is getting kind of weird. There's little certainty, and one thing we can all do in the midst of it is become more self-sufficient. For me, that's raising chickens and growing my own food. For you, that might mean something different.

After a lengthy search for my perfect modern homestead property, I found what is now the Epic Homestead. My 0.3-acre (0.12 ha) oasis in San Diego has allowed me to grow and stretch my capacities as a food grower, a chicken keeper, and a more self-sufficient person. When I had my first failed cucumber experiment back in my apartment years ago, I didn't know I'd end up here. I certainly didn't get here overnight, rather it was step by step, learning a little about everything as I went.

I'm glad I started where I did so I can have an appreciation for what it's like to want to be more self-sufficient while living in a small space. I want to encourage you to do what you can where you are. That might mean you're growing microgreens in the hall closet, installing a rain barrel to water fruit trees in containers on the patio, or keeping a full collection of chickens, quail, and bees.

The great thing about modern homesteading is that the term is so much broader than what is traditionally thought of as "homesteading." We get to use new technologies alongside age-old knowledge to live lighter on the planet, eat great food, have more peace of mind, and come together with people close to us.

Modern homesteading is still a lot of work. You can spend endless time on your pursuit for some level of self-sufficiency. If you're going to find a way to pass your time, though, there are few ways that are better than this.

Here in a climate where I grow year-round, it's easy to experience burnout. In any climate, come August or September, it's really natural to feel a little bit over it. It's so hot, and you might not want to go out to the garden to work. Grow a garden at a scale that's comfortable for you so you experience less of this. Also, there's nothing wrong with taking a break. If you're taking better care of your garden than you are of yourself, it might be a signal that you should let go of something in the garden and come back to it later.

In the hustle of it all, I try to keep a positive outlook. Sometimes this is harder than others. My advice to you: If you lose a harvest that you spent a lot of time on and you're feeling sad about your garden, find something that excites you to start growing again to refresh yourself. Pick something that might be a little easier or that might be fun, even if it's not super productive. Try growing some flowers. Just do something that gets you excited to be in the garden again. Not everything we do has to be in the name of productivity.

It's also been helpful for me to accept that things in the garden don't always look pretty or look the way I think they "should." This is especially hard for me because I invite you into my garden with the Epic Gardening and Epic Homesteading YouTube videos all the time! Of course I want everything to look nice. But consider the giant indeterminate tomato plant I had in one of my first seasons at the Epic Homestead. This plant produced hundreds of tomatoes over its lifetime, and by the end of the summer, it looked terrible. It was still producing more tomatoes than I needed, so I couldn't pull it out just because it was unattractive. I'm not suggesting you let garden pests get out of control or allow a problem to start, but consider where you want to put your resources before you jump on spraying and upending all of the plants that aren't perfect.

What I call food insurance—this homestead investment into my own food supply—was a major motivation to starting my modern homesteading journey. I've come to learn—and hope I was able to demonstrate to you in this book—that homesteading is not just about the food but about our approach to modern living.

Responsible water use—including rainwater catchment and greywater reuse—is huge. Just in the time I've spent writing this book, San Diego has had more rain in a matter of months than we typically see in a couple of years. We have an abundance of water right now, but next year, we could very well be back in drought. This uncertainty only makes me want to be more prepared for whatever is coming next. I can catch water now to use later, and I like that peace of mind.

Energy use, too, is an essential piece of more-sustainable living. The sun is out here offering energy for the taking. It's up to us to make the investment to use our resources wisely. Personally, I'm excited about how renewable-energy technologies are advancing and becoming more available to regular people like us—not just big energy conglomerates.

Homesteading is also about reconnecting with our physical and spiritual selves in a way the typical modern lifestyle has neglected. Participating in my own food production makes me appreciate the food so much more. It feels good to have this level of food security, to act on my environmental concerns, and to spend more time outside and with people I enjoy having in my life.

If, like me, you're looking for more freedom and self-reliance in your life without sacrificing modern comforts, I hope that what you've learned from my journey, from this book, and from the Epic Homesteading and Epic Gardening channels helps you along the way.

Good luck in the garden, and keep on growing.
—Kevin

REFERENCES

Introduction

Barrett, Diane. "Maximizing the Nutritional Value of Fruits & Vegetables," *Food Technology*, Volume 61 (2007) https://fruitandvegetable.ucdavis.edu/files/197179.pdf.

Fischer, Sahrah, Thomas Hilger, Hans-Peter Piepho, Irmgard Jordan, Jeninah Karungi, Erick Towett, Keith Shepherd, Georg Cadisch. "Soil and farm management effects on yield and nutrient concentrations of food crops in East Africa," *Science of the Total Environment*, Volume 716 (2020) https://doi.org/10.1016/j.scitotenv.2020.137078.

Montgomery, David R., and Anne Biklé. "Soil Health and Nutrient Density: Beyond Organic vs. Conventional Farming," *Frontiers in Sustainable Food Systems*, Volume 5 (2021) https://www.frontiersin.org/articles/10.3389/fsufs.2021.699147/full.

CHAPTER 1
Site Selection

"Frost Dates: First and Last Frost Dates by Zip Code," The National Gardening Association (Accessed Nov. 11, 2022) https://garden.org/apps/frost-dates/.

"Homeowners Associations and Urban Ag Law," Sustainable Economies Law Center (Accessed Nov. 11, 2022) https://urbanaglaw.org/homeowners-associations/.

"Planning and Zoning," Sustainable Economies Law Center (Accessed Nov. 11, 2022) https://urbanaglaw.org/planning-and-zoning/.

Slack, Val. "Zoning— What Does It Mean to Your Community?" Purdue University Cooperative Extension Service, ID-233 (2000) https://www.extension.purdue.edu/extmedia/ID/ID-233.pdf.

"USDA Plant Hardiness Zone Map," United States Department of Agriculture (Accessed Nov. 11, 2022) https://planthardiness.ars.usda.gov/.

Ward, Mary. "Growing Zones and Frost Dates - How To Schedule Planting," Gardening.org (March 20, 2022) https://gardening.org/growing-zones-and-frost-dates/.

"Zoning for Urban Agriculture: A Guide for Updating Your City's Laws to Support Healthy Food Production and Access," Healthy Food Policy Project (Accessed Nov. 11, 2022) https://healthyfoodpolicyproject.org/key-issues/zoning-for-urban-agriculture.

CHAPTER 2
Outdoor Food Growing

Chalker-Scott, Linda. "Hugelkultur: What is it, and should it be used in home gardens?" Washington State University Extension (2017) http://pubs.cahnrs.wsu.edu/publications/wp-content/uploads/sites/2/publications/FS283E.pdf.

"Cover Crops, Brassicas," UMass Extension Vegetable Program (Jan. 2013) https://ag.umass.edu/vegetable/fact-sheets/cover-crops-brassicas. Pankau, Ryan. "Keyhole Gardens," Illinois Extension (Sept. 6, 2017) https://extension.illinois.edu/blogs/garden-scoop/2017-09-06-keyhole-gardens.

"Vegetable Production Chart," Michigan State University College of Agriculture and Natural Resources (Accessed Nov. 21, 2022) https://www.canr.msu.edu/uploads/files/Table%204.pdf.

CHAPTER 3
Indoor Food Growing

"DIY Sprouts," University of Florida Institute of Food and Agricultural Sciences Gardening Solutions (Accessed Dec. 5, 2022) https://gardeningsolutions.ifas.ufl.edu/design/outdoor-living/diy-sprouts.html.

"How to hand pollinate tomatoes, peppers and squash?" PlantVillage (2013) https://plantvillage.psu.edu/posts/3422-tomato-how-to-hand-pollinate-tomatoes-peppers-and-squash.

LaBorde, Luke. "What You Should Know About Sprouts," PennState Extension (March 1, 2005) https://extension.psu.edu/what-you-should-know-about-sprouts.

"Mighty Microgreens," University of Maryland Department of Nutrition and Food Science (Sept. 6, 2012) https://nfsc.umd.edu/news/mighty-microgreens.

"The Ultimate Microgreen Cheat Sheet," Bootstrap Farmer (Accessed Dec. 5, 2022) https://www.bootstrapfarmer.com/blogs/microgreens/the-ultimate-microgreen-cheat-sheet.

CHAPTER 4
A Productive Orchard

"Backyard Orchard Culture," Dave Wilson Nursery (Accessed Dec. 12, 2022) https://www.davewilson.com/home-garden/backyard-orchard-culture/.

Bessin, Ric. "Plum Curculio," University of Kentucky College of Agriculture, Food and Environment, Entfact-202 (Aug. 2017) https://entomology.ca.uky.edu/files/efpdf1/ef202.pdf.

Caprile, J.L., and P. M. Vossen. "Pest Notes: Codling Moth," University of California Agriculture and Natural Resources, Publication 7412 (May 2011) https://ipm.ucanr.edu/PMG/PESTNOTES/pn7412.html.

"Cooperative Extension: Tree Fruits," University of Maine Cooperative Extension Tree Fruits Program (Accessed Dec. 12, 2022) https://extension.umaine.edu/fruit/.

Grafton-Cardwell, Elizabeth E., and Matthew P. Daugherty. "Pest Notes: Asian Citrus Psyllid and Huanglongbing Disease," University of California Agriculture and Natural Resources, Publication 74155 (Sept. 2018) https://ipm.ucanr.edu/legacy_assets/pdf/pestnotes/pnasiancitruspsyllid.pdf.

Hoover, Emily E., Emily S. Tepe, Annie Klodd, Marissa Schuh, and Doug Foulk. "Growing Apples in the Home Garden," University of Minnesota Extension (2021) https://extension.umn.edu/fruit/growing-apples.

"How Do I Prune My Backyard Fruit Tree?" Dave Wilson Nursery (Accessed Dec. 12, 2022) https://www.davewilson.com/home-garden/faq/question/how-prune-fruit-tree/.

Koetter, Rebecca, and Michelle Grabowski. "Fire Blight," University of Minnesota Extension (2019) https://extension.umn.edu/plant-diseases/fire-blight.

Lessig, Mack. "Backyard Food Forest," University of Florida Institute of Food and Agricultural Sciences Extension, Manatee County (Sept. 3, 2019) https://blogs.ifas.ufl.edu/manateeco/2019/09/03/backyard-food-forest/.

Lord, William G., and Amy Ouellette. "Growing Fruit: Grafting Fruit Trees in the Home Orchard," University of New Hampshire Cooperative Extension (March 2017) https://extension.unh.edu/sites/default/files/migrated_unmanaged_files/Resource003733_Rep5323.pdf.

Mahr, Sarah. "Espalier," University of Wisconsin – Madison Division of Extension (Accessed Dec. 12, 2022) https://hort.extension.wisc.edu/articles/espalier/.

Polomski, Robert F. "Pruning and Training Apple and Pear Trees," Clemson Cooperative Extension Home and Garden Information Center, Factsheet 1351 (February 2000) https://hgic.clemson.edu/factsheet/pruning-training-apple-pear-trees/.

Schrader, Tom. "Aphids on Apple Trees," The Coastal Gardener: Humboldt & Del Norte Master Gardeners Newsletter, University of California Agriculture and Natural Resources (June 20, 2021) https://ucanr.edu/blogs/blogcore/postdetail.cfm?postnum=47880.
Strang, John. "Fruit and Vegetable Ripening Dates in Kentucky," University of Kentucky Cooperative Extension Service, HortFact-3000 (April 2006) https://www.uky.edu/hort/sites/www.uky.edu.hort/files/documents/ripedate06.pdf.

Thornton, Holly, and Phil Brannen. "Diagnostic Guide to Common Home Orchard Diseases," University of Georgia Extension, Bulletin 1336 (May 2015) https://secure.caes.uga.edu/extension/publications/files/pdf/B%201336_7.pdf.

"What is the Proper Way to Plant a Bare Root Tree," Horticulture and Home Pest News, Iowa State University Extension and Outreach (Accessed Dec. 12, 2022) https://hortnews.extension.iastate.edu/faq/what-proper-way-plant-bare-root-tree.

Williamson, Joey. "Citrus Insects and Related Pests," Clemson Cooperative Extension Home and Garden Information Center, Factsheet 2221 (Aug. 13, 2019) https://hgic.clemson.edu/factsheet/citrus-insects-related-pests/.

Wunderlich, L.R., J.L. Caprile, P.M. Vossen, L.G. Varela, J.A. Grant, H.L. Andris, W.J. Bentley, W.W. Coates, and C. Pickel. "UC IPM Pest Management Guidelines: Apple," University of California Agriculture and Natural Resources, Publication 3432 (October 2015) https://ipm.ucanr.edu/agriculture/apple/stink-bugs/.

Zane, Nadia. "The Importance of Chill Hours for Fruit Trees," What's Growing On - San Joaquin University of California Master Gardeners Blog (Jan. 14, 2015) https://ucanr.edu/blogs/blogcore/postdetail.cfm?postnum=16468.

CHAPTER 5
Composting

"205.203 Soil fertility and crop nutrient management practice standard," Code of Federal Regulations, Organic Foods Production Act Provisions, National Organic Program (Accessed Dec. 24, 2022) https://www.ecfr.gov/current/title-7/subtitle-B/chapter-I/subchapter-M/part-205/subpart-C/section-205.203.

"Compost," University of Florida Institute of Food and Agricultural Sciences Extension (Accessed Dec. 24, 2022) https://sfyl.ifas.ufl.edu/sarasota/natural-resources/waste-reduction/composting/.

"Making and Using Compost," University of Missouri Extension (July 2022) https://extension.missouri.edu/publications/g6956.

Judd, Michael, Charles David Ray, and Serap Gorucu. "Be Safe Around Wood Pallets," PennState Extension (May 12, 2021) https://extension.psu.edu/be-safe-around-wooden-pallets.

"Pallet Markings and What They Mean," General Pallets and Crates (Oct. 17, 2018) https://generalpallets.com/pallet-markings-and-what-they-mean/.

Trautmann, Nancy, and Elaina Olynciw. "Compost Microorganisms," Cornell Waste Management Institute (Accessed Dec. 24, 2022) https://compost.css.cornell.edu/microorg.html.

CHAPTER 6
Energy Systems

Barbose, Galen, Naïm Darghouth, Eric O'Shaughnessy, and Sydney Forrester. "Tracking the Sun, 2022 Edition, Summary Brief," Berkeley Lab (September 2022) https://emp.lbl.gov/sites/default/files/3_tracking_the_sun_2022_summary_brief.pdf.

"Ductless Mini-Split Heat Pumps," U.S. Department of Energy, Office of Energy Saver, Office of Energy Efficiency & Renewable Energy (Accessed Jan. 3, 2023) https://www.energy.gov/energysaver/ductless-mini-split-heat-pumps.

"Guide to Energy Efficient Lighting," U.S. Department of Energy, Energy Efficiency and Renewable Energy Information Center, DOE/EE-0344 (October 2010) https://www.energy.gov/sites/prod/files/guide_to_energy_efficient_lighting.pdf.

"Lighting, Crime, and Safety," International Dark Sky Association (Accessed Jan. 3, 2023) https://www.darksky.org/light-pollution/lighting-crime-and-safety/.

"Light Pollution Wastes Energy and Money," International Dark Sky Association (Accessed Jan. 3, 2023) https://www.darksky.org/light-pollution/energy-waste/.

Meng, Lin, Yuyu Zhou, Miguel O Román, Eleanor C Stokes, Zhuosen Wang, Ghassem R Asrar, Jiafu Mao, Andrew D Richardson, Lianhong Gu, and Yiming Wang. "Artificial light at night: an underappreciated effect on phenology—of deciduous woody plants," *PNAS Nexus*, Volume 1, Issue 2 (May 2022) https://doi.org/10.1093/pnasnexus/pgac046.

"Net Metering," Solar Energy Industries Association (Accessed Jan. 3, 2023) https://www.seia.org/initiatives/net-metering.

Schoessow, Kevin. "Using Wood Ash in the Home Garden," University of Wisconsin - Madison Extension, XHT1268 (Feb. 27, 2020) https://hort.extension.wisc.edu/files/2020/06/Using_Wood_Ash_in_the_Home_Garden.pdf.

"State Solar Energy Renewable Certificate Markets," U.S. Environmental Protection Agency (Aug. 26, 2022) https://www.epa.gov/greenpower/state-solar-renewable-energy-certificate-markets.

Sullins, Ben. "Using Tesla Powerwall 2 to Pay Less than $1/Day for Electricity," @BenSollinsOfficial YouTube (May 12, 2020) https://www.youtube.com/watch?v=dZapMxbsujM.

"What Time of Use Rate Makes Sense for Residential Energy Arbitrage?" Renewable Energy World (May 20, 2020) https://www.renewableenergyworld.com/storage/what-time-of-use-rate-makes-sense-for-residential-energy-arbitrage/.

CHAPTER 7
Water Conservation

"Laundry to Landscape: Greywater System Example," Marin Water (Accessed Jan. 24, 2023) https://www.marinwater.org/sites/default/files/2021-03/L2L%20Graywater%20Illustration%20and%20Equipment%201-4-21.pdf.

"Oasis Biocompatible Cleaners," Bio Pac Cleaning Products (Accessed Jan. 24, 2023) https://www.bio-pac.com/oasis-biocompatible-cleaners/.

"Rainwater Harvesting," Texas A&M AgriLife Extension (Accessed Jan. 24, 2023) https://rainwaterharvesting.tamu.edu/catchment-area/.

"Showerheads," U.S. Environmental Protection Agency (May 23, 2022) https://www.epa.gov/watersense/showerheads.

Springer, Alicia, "'Slow it, Spread it, Sink it'—Creating a Rain Garden in Your Home Landscape," The Real Dirt Blog, University of California Master Gardeners of Butte County (Nov. 2, 2018) https://ucanr.edu/blogs/dirt/index.cfm?tagname=rain%20gardens.

CHAPTER 8
Mini Livestock

Bolshakova, Virginia Lj., and Elina L. Niño. "Bees in the Neighborhood: Best Practices for Urban Beekeepers," University of California Agriculture and Natural Resources, Publication 8596 (May 2018) https://cambp.ucdavis.edu/sites/g/files/dgvnsk2526/files/inline-files/bees-in-the-neighborhood.pdf.

"The Langstroth Hive," Bee Built (Accessed Feb. 7, 2023) https://beebuilt.com/pages/langstroth-hives.

Lewis, Mike. "Integrating Pastured Quail into a Whole-Farm System," National Center for Appropriate Technology (Oct. 26, 2021) https://attra.ncat.org/integrating-pastured-quail-into-a-whole-farm-system/.

Lounsbury Griffin, Shelly. "Planting a Chicken-Friendly Garden," Washington State University Kittitas County Extension Master Gardeners (2021) https://s3.wp.wsu.edu/uploads/sites/2080/2021/02/Planting-a-Chicken-Friendly-Garden.pdf.

Niño, Bernardo. "Hive Types and Equipment," University of California Agriculture and Natural Resources Cooperative Extension, University of California - Davis Department of Entomology, El Niño Bee Lab (Accessed Feb. 7, 2023) https://ucanr.edu/sites/sandiegobees/files/254413.pdf.

"Quail – Alternative Poultry for the Homestead – Part 1," *The Lafayette Homestead*, University of Arkansas Lafayette County Cooperative Extension Office (April 2016) https://www.uaex.uada.edu/counties/lafayette/The%20Lafayette%20Homestead%20April%202016.pdf.

CHAPTER 9
Food Preservation and Storage

Ball Blue Book Guide to Preserving, Ball (2009).

Christian, Candice. "How to Make Quick Refrigerator Pickles," North Carolina Cooperative Extension (Aug. 10, 2021) http://go.ncsu.edu/readext?684156.

"Classic Sauerkraut," Kraut Source (Accessed Feb. 13, 2023) https://www.krautsource.com/blogs/recipes/classic-sauerkraut

Drake, Barbara H. "Selecting, Storing, and Using Fresh Herbs," Ohio State University Extension, HYG-5520 (July 26, 2021) https://ohioline.osu.edu/factsheet/hyg-5520.

"Fermentation," University of California Cooperative Extension Master Food Preserver Program of Orange County (Accessed Feb. 13, 2023) https://ucanr.edu/sites/MFPOC/How_to_Preserve_Food_at_Home/Fermentation/

Garden-Robinson, Julie, and Allie Benson. "How to Select and Store Fruit," North Dakota State University Extension, FN1845 (April 2022) https://www.ndsu.edu/agriculture/sites/default/files/2022-05/fn1845.pdf.

Hazzard, R. "Garlic Harvest, Curing, and Storage," UMass Extension Vegetable Program (Jan. 2013) https://ag.umass.edu/vegetable/fact-sheets/garlic-harvest-curing-storage. Herman, Marilyn, and Suzanne Driessen. "Preserving herbs by freezing or drying," University of Minnesota Extension (2021) https://extension.umn.edu/preserving-and-preparing/preserving-herbs-freezing-or-drying.

Hirneisen, Andy, and Nicole McGeehan. "Let's Preserve: Freeze Drying," PennState Extension (May 20, 2021) https://extension.psu.edu/lets-preserve-freeze-drying.

Huyck, Linda. "The history of preserving food at home," Michigan State University Extension (Feb. 15, 2021) https://www.canr.msu.edu/news/food_preservation_is_as_old_as_mankind.

Janssen, Don. "Potatoes: Harvesting and Storing," University of Nebraska-Lincoln Extension in Lancaster County (Aug. 2022) https://lancaster.unl.edu/hort/articles/2002/potatocare.shtml.

Kemble, Joseph, and Ed Sikora. "Harvesting and Curing Sweet Potatoes," Alabama Cooperative Extension System, ANR-1111 (Sept. 2020) https://www.aces.edu/wp-content/uploads/2019/03/ANR-1111-Harvesting-Curing-Sweet-Potatoes_093020L-A.pdf.

"Liquid Eggs to Dry Conversion Worksheet," American Egg Board (Accessed Feb. 13, 2023) https://incredibleegg.wpenginepowered.com/wp-content/uploads/2021/06/LiquidWholeEggstoDriedWholeEggs.pdf.

Schmutz, Pamela, and Susan Barefoot. "Canning Foods—The Ph Factor," Clemson Cooperative Extension Home and Garden Information Center, Factsheet HGIC 3030 (Aug. 2011) https://hgic.clemson.edu/factsheet/canning-foods-the-ph-factor/.

Tong, Cindy. "Harvesting and storing home garden vegetables," University of Minnesota Extension (2021) https://extension.umn.edu/planting-and-growing-guides/harvesting-and-storing-home-garden-vegetables.

"What is the proper way to harvest and store winter squash?" Horticulture and Home Pest News, Iowa State University Extension and Outreach (Accessed Feb. 13, 2023) https://hortnews.extension.iastate.edu/faq/what-proper-way-harvest-and-store-winter-squash.

ABOUT KEVIN

Kevin Espiritu is the founder of Epic Gardening. As a self-taught gardener, Espiritu has spent over a decade producing educational gardening content across YouTube, TikTok, Instagram, Facebook, The Beet Podcast, and the Epic Gardening blog, all in service of a simple mission: *teach the world to grow*.

Epic Gardening has amassed over 10 million social media followers, 21 million podcast downloads, and 100 million blog visits. In 2022, Kevin acquired leading seed company Botanical Interests to add to Epic Gardening's growing offering of high-quality gardening products and tools.

In 2020, he purchased his first home in San Diego, California, USA: a small 1,000 square-foot (93 m^2) home on a 13,000 square-foot (1,208 m^2) lot. This became his home base for experimenting beyond the garden into energy systems, composting, chicken keeping, canning, preserving, and water capture. These experiences formed the basis for this book, *Epic Homesteading*.

Additionally, Espiritu has authored two other books, *Field Guide to Urban Gardening* and *Grow Bag Gardening*.

INDEX